Studies in Systems, Decision and Control

Volume 410

The series “Studies in Systems, Decision and Control” (SSDC) covers both new developments and advances, as well as the state of the art, in the various areas of broadly perceived systems, decision making and control–quickly, up to date and with a high quality. The intent is to cover the theory, applications, and perspectives on the state of the art and future developments relevant to systems, decision making, control, complex processes and related areas, as embedded in the fields of engineering, computer science, physics, economics, social and life sciences, as well as the paradigms and methodologies behind them. The series contains monographs, textbooks, lecture notes and edited volumes in systems, decision making and control spanning the areas of Cyber-Physical Systems, Autonomous Systems, Sensor Networks, Control Systems, Energy Systems, Automotive Systems, Biological Systems, Vehicular Networking and Connected Vehicles, Aerospace Systems, Automation, Manufacturing, Smart Grids, Nonlinear Systems, Power Systems, Robotics, Social Systems, Economic Systems and other. Of particular value to both the contributors and the readership are the short publication timeframe and the world-wide distribution and exposure which enable both a wide and rapid dissemination of research output.

Indexed by SCOPUS, DBLP, WTI Frankfurt eG, zbMATH, SCImago.

All books published in the series are submitted for consideration in Web of Science.

More information about this series at https://link.springer.com/bookseries/13304

Omar Naifar · Abdellatif Ben Makhlouf
Editors

Advances in Observer Design and Observation for Nonlinear Systems

Fundamentals and Applications

Springer

Editors
Omar Naifar
Department of Electrical Engineering, CEM Laboratory
University of Sfax, National School of Engineering
Sfax, Tunisia

Abdellatif Ben Makhlouf
Department of Mathematics
Jouf University, College of Science
Sakaka, Saudi Arabia

ISSN 2198-4182 ISSN 2198-4190 (electronic)
Studies in Systems, Decision and Control
ISBN 978-3-030-92730-1 ISBN 978-3-030-92731-8 (eBook)
https://doi.org/10.1007/978-3-030-92731-8

This Springer imprint is published by the registered company Springer Nature Switzerland AG
The registered company address is: Gewerbestrasse 11, 6330 Cham, Switzerland

Preface

The challenge of observer design comes naturally in a system approach whenever there is a need for internal knowledge from exterior measurements (directly available). In general, we cannot use as many sensors as there are signals of interest characterizing the system's behavior (due to technological constraints, cost, etc.), especially when such signals can be quite large in number and of various types: they typically include constant signals (parameters), time-varying signals characterizing the system (state variables), and random signals (disturbances). This information is required for a variety of reasons, including modeling system management (control), (identifying), and monitoring (fault detection). All of these goals are required to work together to maintain a regulated system, as seen in Fig. 1 below. As a result, the observer problem is central to a generic control theory problem. The goal of this section will so be to provide a review of various potential techniques for the design of observers (the linked issues of identification, control, and fault detection): In a nutshell, an observer is information, reconstructive, closed-loop, model-based, and measurement-based. In actuality, the model is in the form of state space, and we'll assume that state variables include all of the information that needs to be reconstructed. In this case, we can either construct an explicit dynamic system whose state approximates the true state of the nonlinear model, or we can easily fix the issues through optimization. The first example will be discussed in this article, while the second will be discussed in the following one. In terms of the model under consideration, it might be either discrete time or continuous time, stochastic or deterministic, infinite dimension or finite, "with singularities" or smooth. However, in order to provide a reasonably consistent presentation, this review paper will confine itself to continuous-time state-space descriptions, finite-dimensional, smooth and deterministic (even if some observer design concepts can be described in the literature for several other cases).

In a nutshell, the goal is to focus on specific structures that are available to observers (observer forms, as indicated in the diagram) and try to adapt the system under investigation to such conditions. As a result, many foundations for observer

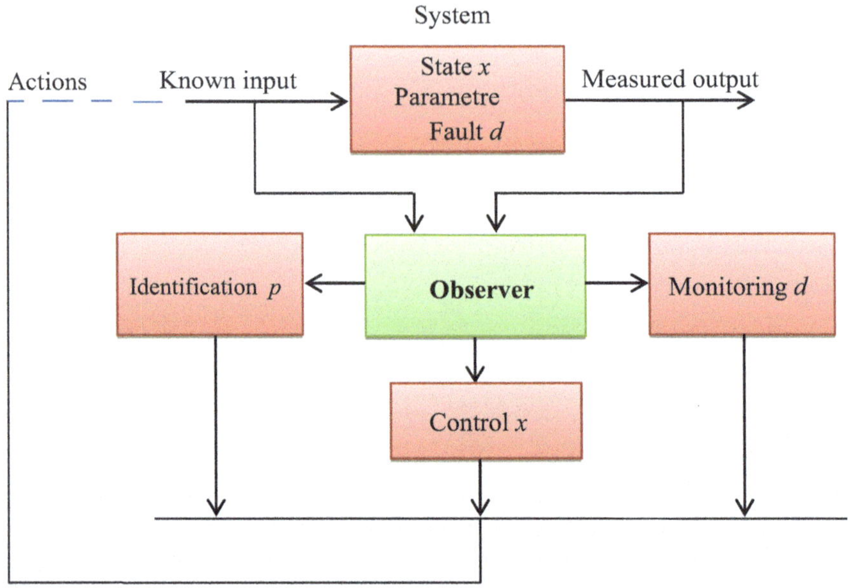

Fig. 1 The observer serves as the control system's heart.

design are presented initially, with two types of corrective gains: steady (input-independent) and time-varying (possibly input-dependent). Then, either through linkages or transformations, techniques for hypothetical expansions of such designs to more generic kinds of systems are given.

This book is composed of nine chapters:

- Chapter "Observers and Observability for Continuous-time Linear Systems"
  The authors introduce the concepts of observability and observers for continuous-time linear systems in this chapter. A general introduction is given first. The authors next look at the observability issue for linear systems. The notion of observability, as well as several observability criteria and certain instances, are discussed. The concept of observers for linear systems is introduced in the final section. The importance of both minimal-order and full-order observers is stressed.
- Chapter "Generalities on Linear Matrix Inequalities and Observer Design of Linear Systems"
  The Linear Matrix Inequalities (LMIs) are discussed in this chapter. There was also discussion of the words "observability" and "observer." A linear observer-based control scheme employing the LMIs approach is also demonstrated.
- Chapter "Some Preliminaries on Unknown Input Observers, Discontinuous Observers and Sliding Mode Observers Design"
  In this chapter, some preliminaries on unknown input observers are given. Furthermore, the development of discontinuous observers and the methods in which they

are designed are introduced. The final parts of the chapter are dedicated to the sliding mode observer's design.

- Chapter "Observers with Unknown Inputs of Linear Systems"
  This chapter introduces the concepts of observability and observer for linear systems, as well as the concept of sliding mode. It is demonstrated how to build an observer for a linear dynamical system with unknown inputs. The determination of the observer's gain in order to ensure convergence leads to the solution of an LMI issue (bilinear matrix inequalities). The technique based on a change of variables is used to resolve these LMI restrictions, allowing the matrices characterizing the observer to be determined.
- Chapter "Luenberger Observer of Impulsive Systems: A Survey"
  In this chapter, some preliminaries on the state estimation of impulsive systems have been conducted. This problem is rarely tackled for this class of systems by researchers and they have designed an observer in the case of autonomous impulsive linear systems, under the condition of strong observability, which does not exist in this work.
- Chapter "Compensator Design Via the Separation Principle for a Class of Nonlinear Uncertain Evolution Equations on a Hilbert Space"
  Several researches have been conducted on the construction of compensators for evolution equations in Hilbert spaces using the separation principle. A non-linear time-varying Luenberger observer has been developed to estimate the system states under uniformly Lipschitz continuous perturbation. The Luenberger observer, which is based on a linear controller, has been shown to stabilize the system. Partial differential equations are used to apply these findings.
- Chapter "Observer Desing for Non Linear Takagi-Sugeno Fuzzy Systems. Application to Fault Tolerant Control."
  The problem of fuzzy fault tolerant control design for systems described by Takagi-Sugeno models is studied in this chapter. The fault tolerant control design requires the state and fault estimation. In order to make this estimation, a proportional integral observer is conceived. The proposed method shows that it is possible to conceive simultaneously the proportional integral observer and the fuzzy fault tolerant control. The cases of system affected by actuator and/or sensor faults are considered. In order to conceive the fault tolerant control strategy for the case of sensor faults, a mathematical transformation is used allowing conceiving an augmented system, in which the initial sensor fault appears as an actuator fault. The fault tolerant control and the proportional integral observer are both conceived considering the augmented state. The noise effect on the state and fault estimation is also minimized in this study, which provides some robustness properties to the proposed control and observer. The fault tolerant control and proportional integral observer design is formulated in terms of linear matrix inequalities (LMI).
- Chapter "On Observer Design of Systems Based on Renewable Energy"
  This chapter compares two observation approaches, the adaptive observer and the interconnected observer, as applied to a wind energy conversion system (WECS)-based induction machine (IM). It has been discovered that the introduced interconnected observer outperforms the standard adaptive one. In contrast, an adaptive

interconnected observer is used in the second half of this chapter for both IM and PMSM-based WECS. This type of observer is robust and adjusts for the impacts of parametric changes.

- Chapter "On Observers-Based Controller Design for Induction Machine"
  This chapter examines the stability of a controlled IM employing both adaptive and interconnected observers. The Lyapunov method is used to validate the recently developed robust control. Under parametric changes, the global stability analysis of the closed-loop system is demonstrated. Finally, simulation results for the proposed observer-based controller architecture for the IM are shown.

Sfax, Tunisia Omar Naifar
Sakaka, Saudi Arabia Abdellatif Ben Makhlouf

Contents

Observers and Observability for Continuous-Time Linear Systems

Assaad Jmal, Jalloul Méliani, and Omar Naifar

Abstract In this chapter, the authors present the notions of observability and observers for continuous-time linear systems. First, a general introduction is given. Then, the authors investigate the observability problem for linear systems. The observability definition, as well as different criteria of observability and some examples are exposed. In the final part, the notion of observers for linear systems is presented. Both cases of minimal-order observers and full-order observers are highlighted.

1 Introduction

In the control theory, every dynamical system has its proper set of physical variables. Within the state-space representation, these variables are actually the system states. Note that, generally speaking, not always all the system states are accessible for measurement, and even if they are measurable, there are some challenging aspects. For example, in many situations, it's not always possible to measure all the states using sensors; in real situations, some states are simply impossible to access using sensors. That's said, there is a natural need to concept "software" sensors that would reconstruct the system states. These software sensors are known as observers.

We can assign the following definition: An observer is a "software" "measuring" technique that allows to reconstruct all the states of a given system, by having a minimum of information about these states. This minimum of information is obtained through a physical sensor. An observer therefore makes it possible to optimize the number of sensors in an industrial application, hence his economic interest in the industry. During the last decades much work, in the control theory, has been

A. Jmal · O. Naifar (✉)
Control and Energy Management Laboratory, University of Sfax, National School of Engineering of Sfax, Sfax, Tunisia

J. Méliani
University of Kairouan, Higher Institute of Apllied Sciences and Technology of Kairouan, Kairouan, Tunisia

O. Naifar and A. Ben Makhlouf (eds.), *Advances in Observer Design and Observation for Nonlinear Systems*, Studies in Systems, Decision and Control 410,
https://doi.org/10.1007/978-3-030-92731-8_1

carried out on the design of observers. One crude way to observe the states of a system is to digitally derive the information measured with sensors. Experience has shown that this method has the disadvantage of giving erroneous results due to noise amplification, caused by measurement imperfections.

However, the task of designing an observer is not systematic. There are some sufficient conditions to be verified, in order to be able to practically design an observer. This problem is known as the observability problem.

From several decades, researchers have focused on solving the observability and observer design problems for linear systems. Kalman-Bucy [1] introduced in 1961 a solution for stochastic linear systems. Their result is currently known by the Kalman filter. This filter also gives good results for deterministic systems. In 1964–1971, Luenberger [2, 3] founded the observer theory which bears his name "Luenberger Observers". His idea is to add to the model put in the canonical companion form (Brunovsky) a correction using the measurement provided by the sensors.

The rest of the chapter is organized as follows. In Sect. 2, the observability problem for continuous-time linear systems is highlighted. The general observability definition is cited, two observability criteria are presented, and the relation between models and observability is explained. Section 3 is dedicated to the observer design problem for continuous-time linear systems. Some basic preliminaries are given, and both cases of minimal-order observers and full-order observers are presented.

2 Observability for Continuous-Time Linear Systems

Let's consider a linear system, described by the following equation, for any time $t \geq t_0$:

$$\begin{cases} \dot{x}(t) = Ax(t) + Bu(t) \\ y(t) = Cx(t) \end{cases} \tag{1}$$

where $x(t)$, $u(t)$ et $y(t)$ are vectors of dimensions n,m et p, that represent respectively the state, the input and the output vectors.

Definition 1 Observability

System (1) is said to be observable, if for an initial time t_0, there exists a constant instant t_1 such that the knowledge of $y(t_0,t_1)$ and $u(t_0,t_1)$ allows to determine, in a unique way, the state $x(t_0)$.

It is possible that this definition only holds for a part of the state vector, which then constitute the observable states of the system. The definition of observability makes no specific assumption about the nature of the input. This property can be interpreted as the capacity of a system to reveal the history of its state vector through that of its outputs. It depends in fact only on matrices A and C.

2.1 Kalman Criterion

The observability matrix O of system (1) is defined as follows:

$$O = \left[CCACA^2\ldots\ldots CA^{(n-1)}\right]^T \tag{2}$$

The observability Kalman criterion says the following:

A necessary and sufficient condition for the observability of system (1) (i.e.: the observability of the pair (*A*, *C*) is the regularity of the observability matrix O. That's said, the observability is guaranteed when the rank of the observability matrix is maximum:

$$rank(O) = n \tag{3}$$

Example 1 Consider the system:

$$\begin{cases} \dot{x}(t) = \begin{pmatrix} 0 & 1 \\ 1 & 3 \end{pmatrix} x(t) + \begin{pmatrix} 0 \\ 1 \end{pmatrix} u(t) \\ y(t) = (0\ 1)x(t) \end{cases}$$

The observability matrix here is:

$$O = [CCA]^T = \begin{pmatrix} 0 & 1 \\ 1 & 3 \end{pmatrix}$$

$rank(O) = 2$, which means that the system is observable.

Example 2 Consider the electronic system [4], represented by Fig. 1.

The chosen state vector is $x = [x_1 x_2] = [\, v_{s1}\ v_{s2} - V_{ref}\,]$. The input is the voltage $u = v_1$ and the output is the voltage $y = v_s$. This choice, as well as the application of the rules of electronics, by considering the operational power amplifiers as perfect, lead to the following state representation:

$$\begin{cases} \dot{x}(t) = \begin{pmatrix} -\frac{1}{R_1C_1} & 0 \\ 0 & -\frac{1}{R_2C_2} \end{pmatrix} x(t) + \begin{pmatrix} \frac{1}{R_1C_1} \\ 0 \end{pmatrix} u(t) \\ y(t) = (-1\ -1)x(t) \end{cases} \tag{4}$$

The observability matrix here is:

$$O = \begin{pmatrix} -1 & -1 \\ \frac{1}{R_1C_1} & \frac{1}{R_2C_2} \end{pmatrix}$$

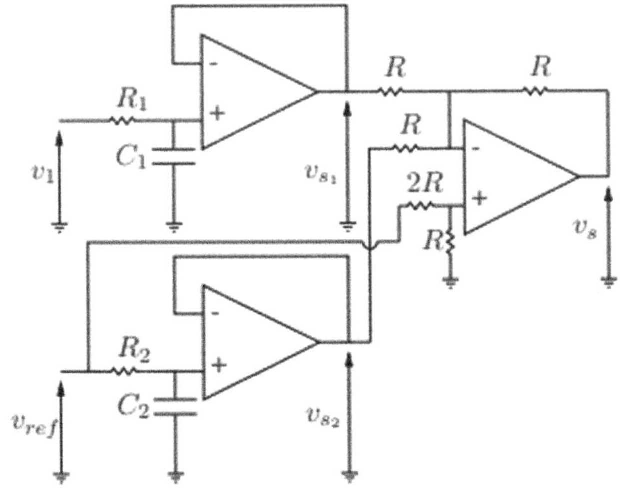

Fig. 1 Electronic system scheme

$rank(O) = 2$, which means that this system is observable.

If someone considers v_{s1} as the output, then the output matrix becomes $C = [10]$ and the observability matrix becomes:

$$O = \begin{pmatrix} 1 & 0 \\ -\frac{1}{R_1 C_1} & 0 \end{pmatrix}$$

It's clear that the rank of this matrix is 1, which means that such a system is no longer observable.

2.2 *Criterion for Jordan Forms: Matrix A Diagonalizable*

In this case, there exists a passage matrix M which makes it possible to change the basis and obtain the following realization:

$$\begin{cases} \dot{x}(t) = \begin{pmatrix} \lambda_1 & 0 & \dots & 0 \\ 0 & \lambda_2 & \dots & 0 \\ \vdots & \vdots & \ddots & \vdots \\ 0 & 0 & \dots & \lambda_n \end{pmatrix} x(t) + \begin{pmatrix} b_1 \\ b_2 \\ \vdots \\ b_n \end{pmatrix} u(t) \\ y(t) = (c_1 \ c_2 \ \dots \ c_n \ ) x(t) \end{cases}$$

In this case, we can assimilate the observability of a mode λ_i to that of component x_i of the associated state vector. Clearly, on this simple diagonal form, we note that we can see an evolution of x_i if the component c_i is not zero.

We say that the mode λ_i *is observable, if and only if* $c_i \neq 0$.

Example 3 If we consider Example 2 of the previous sub-section (the electronic circuit system), we can see from the state-space Eq. (4) that matrix A is diagonal and the elements of the vector C are both non-zero. This means that, according to the Jordan form criterion, the two modes are observable. This result is compatible with the one obtained by the Kalman criterion.

2.3 *Difference Between Models*

Let's consider the composite system given by Fig. 2, made up of three cascade systems [5]:

A state-space representation is first established using x_1, x_2 and x_3 as state variables. The S_3 subsystem clearly shows:

$$\dot{x}_3 = -2x_3 + u$$

Subsystem S_2 gives:

$$\dot{x}_2 = -2x_2 + 3x_2 + v = x_2 - x_3 + u$$

And subsystem S_1 gives the following equation:

$$\dot{x}_1 = -x_1 + 2w = -x_1 + 2x_2 - 2x_3 + 2u$$

These three differential equations allow, by expressing $y = x_1 + w$ in parallel, to write the following state-space representation:

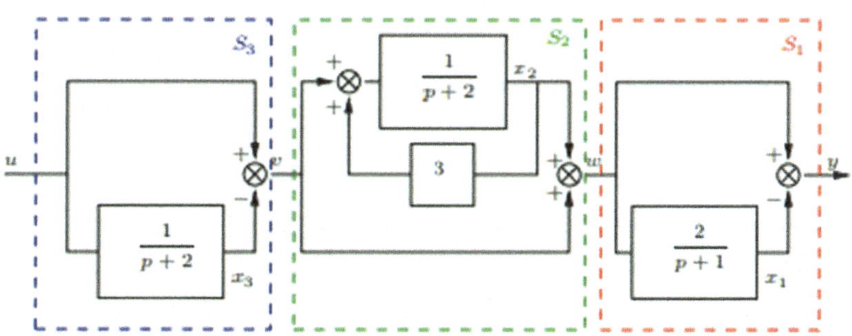

Fig. 2 Composite System

$$\begin{cases} \dot{x}(t) = \begin{pmatrix} -1 & 2 & -2 \\ 0 & 1 & -1 \\ 0 & 0 & -2 \end{pmatrix} x(t) + \begin{pmatrix} 2 \\ 1 \\ 1 \end{pmatrix} u(t) \\ y(t) = (\,-1\ 1\ -1\,)x(t) \end{cases}$$

The system is therefore of order 3 and has for modes $\{-1;\ 1;-2\}$. The observability matrix is:

$$O = \begin{pmatrix} -1 & 1 & -1 \\ 1 & -1 & 3 \\ -1 & 1 & 7 \end{pmatrix}$$

The rank of this matrix is 2, which means that such a system has 1 non-observable mode. Indeed, the rank deficiency provides information on the number of non-observable modes. To know this non-observable mode, we can diagonalize the above state-space equation. We use the following passage matrix:

$$V = \begin{pmatrix} 1 & 1 & 4 \\ 0 & 1 & 1 \\ 0 & 0 & 3 \end{pmatrix}$$

One possible diagonal realization is the following one:

$$\Lambda = V^{-1}AV = \begin{pmatrix} -1 & 0 & 0 \\ 0 & 1 & 0 \\ 0 & 0 & 2 \end{pmatrix}, \tilde{C} = CV = (\,-1\ 0\ -6\,)$$

This diagonal realization, according to the criterion of paragraph 2.2, shows that the mode 1 is not observable.

We now seek to establish the unique differential equation between u and y. The subsystem S_1 corresponds in fact to a block:

$$S_1(p) = \frac{p-1}{p+1}$$

Which gives the following temporal equation:

$$\dot{y} + y = \dot{w} - w$$

Sub-system S_2 corresponds to the following bloc:

$$S_2(p) = \frac{p}{p-1}$$

Which gives the following temporal equation:

$$\dot{w} - w = \dot{v}$$

Finally, sub-system S_3 corresponds to the following bloc:

$$S_3(p) = \frac{p+1}{p+2}$$

Which gives the following temporal equation:

$$\dot{v} + 2v = \dot{u} + u$$

These 3 differential equations lead to the following one:

$$\ddot{y} + 3\dot{y} + 2y = \ddot{u} + \dot{u}$$

This is a differential equation of order, 2 which suggests that the system is of order 2. The associated modes are -1 and -2. The global transfer function of this system is:

$$G(p) = S_1(p)S_2(p)S_3(p) = \frac{p}{p+2}$$

This transfer function lets us assume that the system is of order 1 and that its only mode is -2. So what happened? The differential equation has lost the unobservable mode.

The differential equation represents only the observable part of the system.

We can see here that the state-space equation is a more complete model than the differential equation, which is itself a more complete model than the transfer function.

3 Observers for Continuous-Time Linear Systems

3.1 Preliminaries

3.1.1 Observation Principle

The principle of the observation is to use u and y to reconstruct a vector $\hat{x}$ which is as close as possible to x, in order to then perform a state feedback, as shown in Fig. 3.

As shown in Fig. 3, the observer (also called state reconstructor) together with the state feedback K constitutes a dynamic feedback which has two entries: u and y.

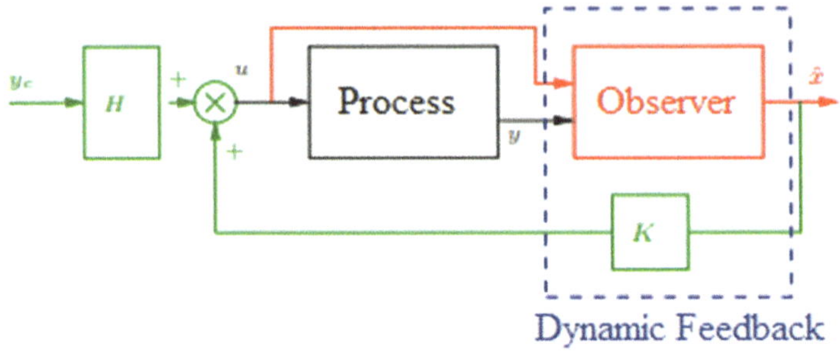

Fig. 3 The Observer Principle

Synthesizing an observer consists of determining, on the basis of the process state model, a state model for the observer. There are several techniques for carrying out this synthesis, but before presenting two of them, we will dwell a little on a few aspects first.

3.1.2 Design Goal of an Observer

The logic of observation is simple. It is utopian to want to construct an observer such that $\hat{x}(t) = x(t)$ for any time t. Indeed, this would mean that the observer reacts infinitely fast to a change in the state of the process even when $\hat{x}(0) \neq x(0)$. On the other hand, one can hope to obtain this equality as $t \to \infty$. So if we define the difference:

$$e(t) = \hat{x}(t) - x(t) \tag{5}$$

The main design goal of an observer would then be:

$$\lim_{t\to\infty} e(t) = 0 \tag{6}$$

3.1.3 Existence Condition of an Observer

A necessary and sufficient condition for the observability of a system is the observability of the pair *(A, C)*.

One can design a state observer, if and only if the pair (A, C) is observable.

3.2 Design of Minimal-Order Observers

In this part, it is a question of synthesizing a minimal order observer. We first give its definition and structure. This is an observer whose state model corresponds to a state vector of minimal dimension. It can be shown that to properly observe a state vector of dimension n, the dimension of the state vector of the observer must be at least $(n-1)$. This leads to an observer model of the form:

$$\begin{cases} \dot{z}(t) = Fz(t) + Py(t) + Ru(t) \\ \hat{x}(t) = Lz(t) + Qy(t) \end{cases} \tag{7}$$

where $z(t) \in \mathbb{R}^{n-1}$. To synthesize an observer here consists in determining suitably $F \in \mathbb{R}^{(n-1)\times(n-1)}$, $L \in \mathbb{R}^{n\times(n-1)}$, $\{P, R\} \in \{\mathbb{R}^{n-1}\}^2$ and $Q \in \mathbb{R}^{n\times p}$, that is, to make sure that Eq. (6) is verified.

If we set the following relation:

$$z(t) = Tx(t), T \in \mathbb{R}^{(n-1)\times n}$$

For every time t, then one can have:

$$\hat{x}(t) = Lz(t) + Qy(t) = LTx(t) + QCx(t) = (LT + QC)x(t)$$

which, since we want to verify $\hat{x}(t) = x(t)$, would lead to:

$$LT + QC = I \tag{8}$$

But, as we have seen, it is utopian to hope for $\hat{x}(t) = x(t), \forall t$ (i.e. here $z(t) = Tx(t)$). Indeed, it does not make sense to want to satisfy the equality $z(t) = Tx(t), \forall t \geq 0$ since $z(t)$ is not a priori equal to a $Tx(0)$. So we just try to get:

$$z(t) = Tx(t) + \mu(t), \, with \lim_{t\to\infty} \mu(t) = 0 \tag{9}$$

Then, the dynamics of $\mu(t)$ would be governed by the following equation:

$$\dot{\mu}(t) = \dot{z}(t) - T\dot{x}(t) = F\mu(t) + (FT - TA + PC)x(t) + (R - TB)u(t)$$

Therefore, we can impose the constraints:

$TA - FT = PC$ and $R = TB$

Thus, we obtain:

$$\dot{\mu}(t) = F\mu(t)$$

which means that (9) holds when F is stable in the sense of Hurwitz, i.e. has only eigenvalues with a real negative part. By taking into account (9) and the second equation of (7), it comes:

$$\hat{x}(t) = (LT + QC)x(t) + L\mu(t)$$

Which, if we consider (8), gives $\hat{x}(t) \to x(t)$ as $t \to \infty$.

Finally, it is by a judicious choice of the model (7) that we succeed in satisfying (9). More precisely, this consists of:

- *Choosing a matrix F, which is Hurwitz stable and whose modes are fater then the ones of matrix A (to be able to observe $x(t)$ faster than it evolves).*
- *Choosing matrices P and T such that: $TA - PC = FT$.*
- *Computing $R = TB$.*
- *Computing L and Q such that $LT + QC = I$.*

3.3 Design of Full-Order Observers

In this part, we consider that the observer is of the same order as the process, ie *n*. The definition and structure of such an observer are presented. A classical structure of such an observer consists in expressing the latter as a loop system as follows:

$$\begin{cases} \dot{\hat{x}}(t) = A\hat{x}(t) + Bu(t) + Z(\hat{y}(t) - y(t)) \\ \hat{y}(t) = C\hat{x}(t) \end{cases} \tag{10}$$

Here, the state vector of the observer is directly $\hat{x}(t)$ and its output $\hat{y}(t)$, are only used to ensure a feedback within the observer itself as shown in Fig. 4.

The model of the observer is clearly based on that of the system and it is in fact a question of determining Z of such so that the dynamic of the observer is faster than

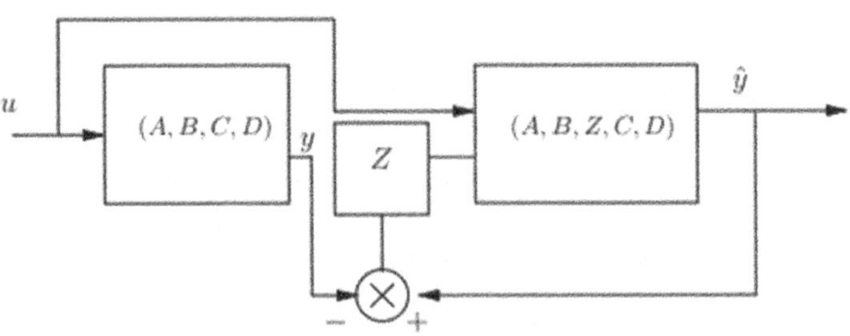

Fig. 4 The full-order observer scheme

that of the process on the one hand, and, on the other hand, that the relation (6) is satisfied. Indeed, the observer Eq. (10) can be rewritten as follows:

$$\dot{\hat{x}}(t) = (A + ZC)\hat{x}(t) + Bu(t) - Zy(t) \tag{11}$$

where we can clearly see that the choice of Z fixes the dynamics of the state matrix $A + ZC$ of the observer. Thus, if we analyze the evolution of the state estimation error $e(t)$, we see that:

$$\dot{e}(t) = (A + ZC)e(t) \tag{12}$$

It is therefore necessary that matrix $A + ZC$ be stable in the sense of Hurwitz, i.e. have all its eigenvalues with a negative real part, so that we have $\hat{x}(t) = x(t)$ as $t \to \infty$. Otherwise (an eigenvalue with a real part that is not strictly negative), $e(t)$ cannot tend towards zero.

4 Conclusion

In this chapter, the authors have presented a theoretical overview of the observability and observer design problem for continuous-time linear systems. First, the authors have investigated the observability problem for linear systems. The observability definition, as well as different criteria of observability and some examples have been exposed. In the final part, the notion of observers for linear systems has been presented. Both cases of minimal-order observers and full-order observers have been highlighted.

References

1. Kalman, R.E., Bucy, R.S.: New results in linear filtering and prediction theory. J. Basic. Eng. **1961**, 95–108 (1961)
2. Luenberger, D.G.: Observing the state of a linear system. IEEE Trans. Military Electron. **8**(2), 74–80 (1964)
3. Luenberger, D.: An introduction to observers. IEEE Trans. Autom. Control **16**(6), 596–602 (1971)
4. Lienig, J.: Bruemmer, Hans, Fundamentals of Electronic Systems Design. Springer (2017). https://doi.org/10.1007/978-3-319-55840-0
5. Bachelier, O.: Cours d'Automatique Représentations d'état linéaires des systèmes monovariables. Link : urlr.me/wbcFd

References

Generalities on Linear Matrix Inequalities and Observer Design of Linear Systems

Intissar Darwich, Dorsaf Etlili, Ayachi Errachdi, and Omar Naifar

Abstract In this chapter, the Linear Matrix Inequalities (LMIs) are presented. The terms "observability" and "observer" were also discussed. A design of linear observer-based control using the technique of LMIs is described, too.

1 Introduction

Over the last two decades, the performance of industrial equipment has improved considerably. The integration of high-performance computers and automation systems makes it possible to develop sophisticated algorithms at the level of control and data processing. However, if the information they use is wrong, these algorithms become ineffective. In the latter situation, the performance of the system is degraded, but to make matters worse, the consequences of the equipment, the environment and the safety of the personnel can be dramatic.

Two points play an important role in automatic. The first is the stability of the system to be studied. In general, the stability is analyzed from its model which can be linear or nonlinear. But as physical systems are never perfectly modeled, so there is uncertainty about its structure (dimensions) and its parameters (numerical values). It is therefore appropriate to have study techniques which take these uncertainties into account. The second issue is the reconstruction of all or part of the state of a system using observers. This is important in two areas of Automation: the control of a system and its monitoring, both of which use knowledge of the system state. Many research are conducted in this area, see [1–15]

The reconstruction of the state of an uncertain system (and by extension the reconstruction of its release) is a classic problem of Automatic. Luenberger [16] studied a state reconstructor which his name has been assigned. The Luenberger

I. Darwich · O. Naifar (✉)
Control and Energy Management Laboratory, University of Sfax, National School of Engineering of Sfax, Sfax, Tunisia

D. Etlili · A. Errachdi
University of Tunis El Manar, National School of Engineering of Tunis, Tunis, Tunisia

O. Naifar and A. Ben Makhlouf (eds.), *Advances in Observer Design and Observation for Nonlinear Systems*, Studies in Systems, Decision and Control 410,
https://doi.org/10.1007/978-3-030-92731-8_2

observer is not always sufficient for fault detection, because the estimation error (of the state or of the output) generated by this observer for an uncertain system or with unknown inputs does not necessarily converge towards zero. In order to overcome this problem, we can use observers of singular systems [17] or observers with unknown inputs [18]. The design problem of proportional and integral action observers, for linear systems with unknown inputs, has also been considered [19, 20].

This chapter discusses Linear Matrix Inequalities (LMIs). There was also debate about the terms "observability" and "observer." The LMIs method is also used to define a linear observer-based control system.

2 Convex Analysis and Linear Matrix Inequalities

2.1 Convex Analysis

The notion of convexity sustains an important place in this book given the chosen orientations. Indeed, the analysis and synthesis problems are formulated, when possible, in terms of convex optimization [21].

The convexity of an optimization problem has a dual benefit:

- the calculation times are reasonable to find a solution;
- there is no local minimum of the cost function to be optimized; the result obtained corresponds to a single global minimum.

The convexity is a concept both set and functional, the following are definitions in each case.

Definition 1 (*Convex set*) Let $\epsilon \subset \mathbb{R}^n$ be a set,ϵ is a convex set if and only if:

$$\forall\lambda \in [01] \subset \mathbb{R}, \forall(x_1, x_2) \in \epsilon^2, \lambda x_1 + (1-\lambda)x_2 \in \epsilon \tag{1}$$

Definition 2 (*Convex function*) Let f be a function $f : \epsilon \subset \mathbb{R}^n \to \mathbb{R}$ with ϵ a convex set, then f is convex if and only if:

$$\forall\lambda \in [01] \subset \mathbb{R}, \forall(x_1, x_2) \in \epsilon^2, f(\lambda x_1 + (1-\lambda)x_2) \leq \lambda f(x_1) + (1-\lambda)f(x_2) \tag{2}$$

Thus, a convex optimization problem is stated as follows: $\min_{x \in e} f(x)$, where ϵ is a convex set and f is a convex function.

Likewise, a constraint $f_i(x) \leq 0$ is known us to be convex if the function f_i is convex. One of the advantages of convexity is that any optimization of a convex function defined on a convex set can be processed locally because any local solution becomes global.

2.2 Classic LMI Problems

In recent years, many studies have emerged with the main objective of reducing a wide variety of synthesis or analysis problems to convex optimization problems involving LMIs. At the same time, efficient methods of solving convex optimization problems have been developed. These methods, called interior-point methods, initially developed by Karmarkar [22] for linear programming, were later extended by Nesterov and Nemirovskii [23] to the case of convex programming in the space of positive definite matrices.

Definition 3 Considering a family of symmetric matrices P_0 and P_i,$i \in \{1, \dots, n\}$ of $\mathbb{R}^{p \times p}$ and a vector $x = (x_1, x_2, \dots x_n)^T \in \mathbb{R}^n$, a strict LMI (respectively not-strict) in x_i, $i \in \{1, \dots, n\}$ is written in this form:

$$F(x) = P_0 + \sum_{i=1}^{n} x_i P_i (\text{respectively} \geq 0) \tag{3}$$

It should be noted that the set E defined by $E = \{x \in \mathbb{R} : F(x) > 0\}$ is convex, which leads us to consider an LMI constraint as a convex constraint.

The form of LMI come across three most common convex optimization which are:

- Realizability (Feasibility) problem: it is a question of finding a vector $\mathcal{S}$ such that the convex constraint $F(x) > 0$ is satisfied. This problem can be solved by finding the vector x minimizing the scalar t such that:

$$-F(x) < t.I \tag{4}$$

  If the minimum value of t is negative, the problem is feasible.
- Eigenvalue problem (EVP): it is a question of minimizing the largest eigenvalue of a symmetric matrix under an LMI type constraint:

$$\begin{aligned} &\text{minimize } \lambda \\ &\text{under the constraints} \begin{cases} \lambda I - A(x) > 0 \\ B(x) > 0 \end{cases} \end{aligned} \tag{5}$$

- Generalized eigenvalue problem (GEVP): this is to minimize the largest generalized eigenvalue of a pair of matrices, with respect to an LMI constraint:

$$\begin{aligned} &\text{minimize } \lambda \\ &\text{under the constraints} \begin{cases} \lambda B(x) - A(x) > 0 \\ \quad B(x) \qquad > 0 \\ \quad C(x) \qquad > 0 \end{cases} \end{aligned} \tag{6}$$

These convex optimization problems can then be solved by different types of methods [24, 25]:

- Secant planes method
- Ellipsoid method
- Simplex type method
- Interior point method

2.3 Writing Constraints in LMI Form

Among the most classic examples of an LMI type constraint we will need throughout this paper it's includes:

- Schur's complement: let three matrices $R(x) = R^T(x)$, $Q(x) = Q^T(x)$ and $S(x)$ be affine with respect to the variable x. The following LMIs are equivalent:

  1. $\begin{pmatrix} Q(x) & S(x) \\ S^T(x) & R(x) \end{pmatrix} > 0$
  2. $R(x) > 0,\ Q(x) - S(x)R^{-1}(x)S^T(x) > 0$

- Convex quadratic constraints: the constraint on the norm $\|Z(x)\| < 1$, where $Z(x) \in \mathbb{R}^{p\times q}$ is affine with respect to the variable $x \in \mathbb{R}^p$ is represented by:

$$\begin{pmatrix} I_p & Z(x) \\ Z^T(x) & I_q \end{pmatrix} > 0 \tag{7}$$

2.4 LMI Regions

Definition 4 Chilali and Gabinet [26]: a region S of the complex plane is named an LMI region if there exists a symmetric matrix $\alpha \in \mathbb{R}^{m\times m}$ ∈ Rm × m is a matrix $\beta \in \mathbb{R}^{m\times m}$ such that:

$$S = \{z \in C : f_S(z) < 0\} \tag{8}$$

with $f_S(z) = \alpha + z\beta + z^*$. The notation z^* signifies the conjugate of z. $f_s(z)$ is so-called the characteristic function of S.

In other words, the LMI region is a region of the complex plane which is characterized by an LMI as a function of z and z^*, or of $a = Re(z)$ and $b = Im(z)$. The LMI regions are therefore convex sets.

Examples of LMI regions.

By setting $a = Re(z)$ and $b = Im(z)$, we get:

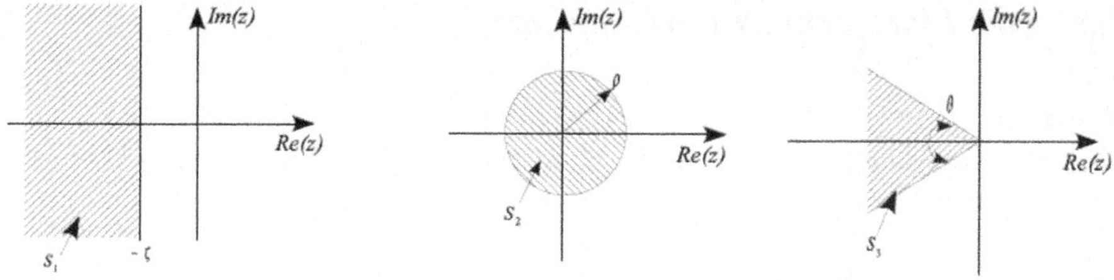

Fig. 1 Examples of LMI regions

$$a = \frac{z^* + z}{2} \text{ and } b = \frac{z - z^*}{2j} \tag{9}$$

The left half plane can be characterized by $a < 0$, the characteristic function of this one is given by:

$$f_S(z) = z^* + z \tag{10}$$

Consider the three regions of the left complex half-plane illustrated in Fig. 1. The region S_1 of the complex plane, $a < -\zeta$, is the LMI region characterized by the following function $f_{S_1}(z)$:

$$f_{S_1}(z) = z^* + z + 2\zeta \tag{11}$$

The disk centered at the origin of the complex plane S_2 is a region characterized by the succeeding relationship:

$$z^* z - \rho^2 < 0 \tag{12}$$

Or by the way of using Schur's complement:

$$f_{S_2}(z) = \begin{pmatrix} -\rho & z \\ z^* & -\rho \end{pmatrix} \tag{13}$$

SectorS_2, $a\tan(\theta) < -|b|$, of the complex plane is an LMI region characterized by the following function $f_{S_3}(z)$ (using Schur's complement):

$$f_{S_2}(z) = \begin{pmatrix} \sin\theta(z + z^*) & \cos\theta(z - z^*) \\ \cos\theta(z^* - z) & \sin\theta(z + z^*) \end{pmatrix} \tag{14}$$

2.5 Pole Placement by LMI Approach

Theorem 1
[26]: the eigenvalues of a real matrix M are placed in an LMI region S (31) of the complex plane if, and only if, there exists a symmetric matrix X such that:

$$M_s(M, X) = \alpha \otimes X + \beta \otimes MX + \beta^T \otimes XM^T < 0 \tag{15}$$

where "$\otimes$" denotes the Kronecker matrix product.

In other words, the eigenvalues of a real matrix M are all in a region of the complex plane, if there is a matrix $X > 0$ such that the LMI $M_S(M, X) < 0$ is achievable, where $M_S(M, X)$ is resolute by performing the following substitution in the characteristic function S:

$$\exists X > 0 : 2\varsigma X + MX + XM^T < 0 \tag{16}$$

The eigenvalues of the matrix M are so all in the region S_1 of the complex plane if and only if:

$$\exists X > 0 : 2\varsigma X + MX + XM^T < 0 \tag{17}$$

Similarly, the eigenvalues of the matrix M are all in the S_2 region of the complex plane if and only if:

$$\exists X > 0 : \begin{pmatrix} -\rho X & MX \\ XM^T & -\rho X \end{pmatrix} < 0 \tag{18}$$

At last, the eigenvalues of the matrix M are all in the region S_3 of the complex plane if and only if:

$$\exists X > 0 : \begin{pmatrix} \sin\theta(MX + XM^T) & \cos\theta(MX - XM^T) \\ \cos\theta(XM^T - MX) & \sin\theta(MX + XM^T) \end{pmatrix} < 0 \tag{19}$$

Theorem 2 Chilali and Gabinet [26]: let be two LMI regions S_1 and S_2 of the complex plane. The eigenvalues of the matrix M are all in the region LMI $S_1 \cap S_2$ if and only if there exists a symmetric matrix $X > 0$ solution of the system.

$$\begin{aligned} M_{S_1}(M, X) < 0 \\ M_{S_2}(M, X) < 0 \end{aligned} \tag{20}$$

These results will be used in the fourth chapter within the framework of the synthesis of the observers gains with unknown inputs.

3 Observability and Observers

The observability of a process is a very important concept in Automatics. Indeed, to reconstruct the state and the output of a system, it is necessary to know, a priori, whether the state variables are observable or not. In general, for reasons of technical feasibility, cost, etc. The size of the output vector is smaller than that of the state. This result drives to at the specified instant$t$, the state $x(t)$ can't be algebraically deduced from the output $y(t)$ at this instant. On the other hand, under observability conditions which will be explained later, this state can be deduced from the knowledge of the inputs and outputs over a previous time interval $([0, t])$, $y([0, t])$.

The goal of an observer is to accurately provide an estimate of the current value of the state based on the previous inputs and outputs. This estimate should be obtained in real time, the observer usually takes the form of a dynamic system.

Definition 5 [27]: an observer of a dynamic system is as follows:

$$\mathcal{S}(t) : \begin{cases} \dot{x}(t) = f(x(t), u(t)) \\ y(t) = h(x(t)) \end{cases} \tag{21}$$

An auxiliary dynamic system $\mathcal{O}$ whose the inputs are the input and output vectors of the system to be observed and whose the output vector $\widehat{x}(t)$ is the estimated state:

$$\mathcal{O}(t) : \begin{cases} \dot{z}(t) = \widehat{f}(z(t), u(t), y(t)) \\ \widehat{x}(t) = \widehat{h}(z(t), u(t), y(t)) \end{cases} \tag{22}$$

where the error between the state vector $x(t)$ and $\widehat{x}(t)$ tends asymptotically towards zero.

$$\|e(t)\| = \|x(t) - \widehat{x}(t)\| \rightarrow 0 while t \rightarrow \infty \tag{23}$$

The diagram of such observer is referred in Fig. (2).

The question asked before any observer synthesis, whether or not its design is possible. The notion of observability and certain properties of the inputs applied to

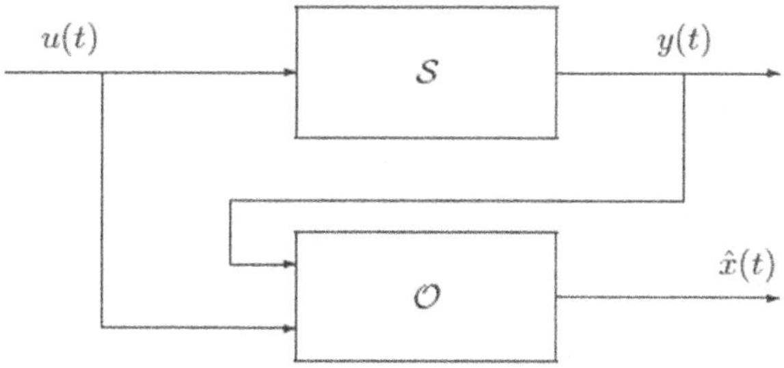

Fig. 2 Structural diagram

the system provide the necessary conditions for the synthesis of an observer. We discuss in this part the observability of linear systems.

3.1 Observability

The fundamental problem of the observability analysis of a physical system is to be able to say whether the state of the system can be determined by inputs and outputs. If that is the case, estimation theory provides tools to reconstruct this state; we recall that the knowledge of unmeasured state components which is generally necessary to tune or to detect the faults of a system.

The initial value of a system's state is generally unknown. We can then request: under what conditions can the state of the system be determined from the outputs and inputs? This issue is named the observability problem. A definition of observability based on the notion of in distinguishability is offered.

Definition 6 [27]: for the system (20), two states x_0 and x_0' are supposed to be indistinguishable, whether for any input function $u(t)$ and for any $t \geq 0$, the outputs $h(x(t), x_0)$ and $h(x(t), x_0')$ (h (x (t)) which result are equal.

Definition 7 [27]: the system (20) is said to be observable if it doesn't have a pair of distinct initial states $\{x_0, x_0'\}$ indistinguishable.

3.2 Observability of Linear Systems

The observability criteria of a linear system are described in numerous references [28–30]. We will only present those concerning certain and regular linear systems. let's us Consider the linear dynamic system:

$$\dot{x}(t) = Ax(t) + Bu(t) \tag{24a}$$

$$y(t) = Cx(t) \tag{24b}$$

where $x(t) \in \mathbb{R}^n$, u(t) $\in \mathbb{R}^m$ and y(t)$\in \mathbb{R}^p$. The matrices A, B and C have appropriate dimensions. The system observability matrix is defined [29] by:

$$\mathcal{O} = \begin{pmatrix} C \\ CA \\ \vdots \\ CA^{n-1} \end{pmatrix} \tag{25}$$

The observability of the system (24) is guaranteed if the rank of the observability matrix $\mathcal{O}$ is equal to n [31]. O'Reilly [28] presented a second criterion; the system (24) is completely observable if:

$$rank = \begin{pmatrix} sI - A \\ C \end{pmatrix} = n \tag{26}$$

for any complex s. If a linear system is completely observable, it is globally observable, that is mean all components of the state vector of the system are observable, and therefore can be reconstructed by an observer. If the system is nonlinear, we need to distinguish global observability from the local one.

3.3 Observer Synthesis for a Certain Linear System

An observer is used for the purpose of estimating state or a linear state function (such as the output of a system) [32, 33]. The comparison of the measured output with its estimated allows to generate signals called "residues" which must be able to inform us about the operating state of the sensors and actuators including the state of the process.

As shown in Fig. (2), a state reconstructor or estimator is a system having as inputs, inputs and outputs of the process and whose output is an estimate of the state of this process. Therefore, we seek to estimate the state of a system linear deterministic defined by (24).

The construction basis of an observer is to correct the estimation error between the actual output and the reconstructed output. This observer is defined by:

$$\dot{\hat{x}} = A\hat{x}(t) + Bu(t) + K\big(y(t) - \hat{y}(t)\big)$$

$$= (A - KC)\hat{x}(t) + Bu(t) + Ky(t)$$

$$\hat{y}(t) = C\hat{x}(t) \tag{27}$$

where $K \in \mathbb{R}^{n \times p}$ is the gain of the observer (27). Having regard to the state and output equations of the observer (27) and the system (24), we deduce the structural diagram presented in Fig. 3.

The observer is synthesized in such a way that the difference between the state of the system and its estimate tends towards zero when t t tends towards ∞, so if the eigenvalues of $(A - KC)$ are in the left half-plane of the complex plane then the gain of the observer K can be determined by the pole placement method if the following theorem is proved:

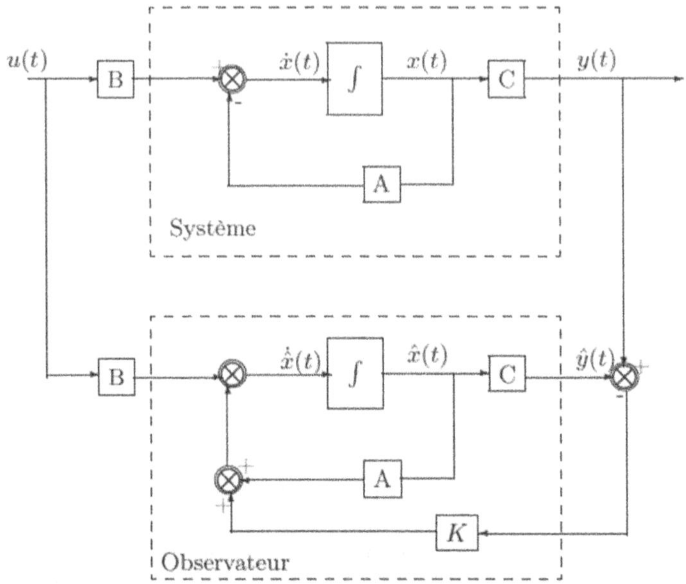

Fig. 3 Structural diagram

Theorem 3 The eigenvalues of $(A - KC)$ can be fixed arbitrarily if and only if the pair (A, C) is observable [32].

If the pair (A, C) is observable, then a boundless freedom allows the user to fix the matrix K. In general, it is chosen such that the eigenvalues of the matrix $(A - KC)$ are in the left half-plane of the complex plane and that the real part of the eigenvalues is greater, in absolute value, than the real part of the eigenvalues of the state matrix A; under these conditions, the dynamics of the observation error is consequently faster than that of the process (system). But there are two contradictory considerations which one must take into account and which interfere in the choice of the matrix K [34]:

- The disturbances on the pair (A, B) lead to, if they are significant, choose a high value of the matrix K in order to enhance the influence of the measurements on the state estimation.
- The noise affecting the measurement of the output quantities, amplified by the gain, requires a small value of K.

Therefore, the observer's gain must be chosen by making a compromise to best meet these constraints.

4 Observer-Based Control Via LMI Approach

In some cases, a part of the state is not accessible. We must then use an observer. We obtain an augmented model composed of a model, a controller and an observer. Such a model is represented on the following Fig. 4:

Consider the stationary linear dynamic system described by:

$$\begin{cases} \dot{x}(t) = Ax(t) + Bu(t) \\ y(t) = Cx(t) \\ x(0) = x_0 \end{cases} \tag{28}$$

where x and u are the state and control vectors respectively. A, B and C are known real matrices.

The corrector with observer is given by:

$$u(t) = K\hat{x}(t) \tag{29}$$

K is the static gain of the corrector and $\hat{x}$ is the estimated state whose dynamics is described by:

$$\dot{x}(t) = A\hat{x}(t) + Bu(t) + L[C\hat{x}(t) - y(t)] \tag{30}$$

L is the gain of the observer.

The stabilization conditions of the system controlled by state feedback from an observer are given by:

$$\begin{aligned} &\begin{bmatrix} J_1 & PB \\ B^T P & -I \end{bmatrix} \prec 0 \\ &\begin{bmatrix} J_2 & K^T \\ K & -I \end{bmatrix} \prec\sim 0 \end{aligned} \tag{31}$$

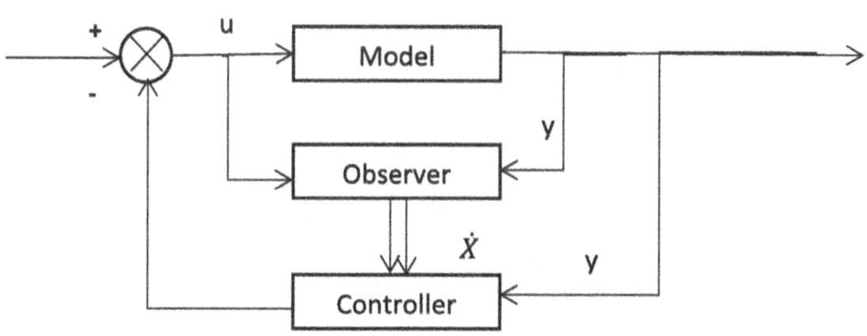

Fig. 4 Representation of the augmented system

With $\begin{array}{l} J_1 = A^T P + PA + K^T B^T P + PBK \\ J_2 = A^T Q + QA + C^T L^T Q + QLC \end{array}.$

Proof Consider the observation error defined by: $e(t) = x(t) - x(t)$.

Replacing $e(t)$ with its value in the observer's equation of state, we obtain:

$$\begin{aligned} \dot{\hat{x}}(t) &= A\hat{x}(t) + Bu(t) + L[C\hat{x}(t) - y(t)] \\ &= A\hat{x}(t) + BK\hat{x}(t) + LC(\hat{x}(t) - x(t)) \\ &= [A + BK]\hat{x}(t) - LCe(t) \end{aligned} \tag{32}$$

$$\begin{aligned} \dot{x}(t) &= Ax(t) + Bu(t) \\ &= Ax(t) + BK\hat{x}(t) \\ &= Ax(t) + BKx(t) - BKx(t) + BK\hat{x}(t) \\ &= [A + BK]x(t) - BK(x(t) - \hat{x}(t)) \\ &= [A + BK]x(t) - BKe(t) \end{aligned} \tag{33}$$

The derivative of the error is then equal to:

$$\begin{aligned} & x(t) - \dot{\hat{x}}(t) = Ax(t) + BK\hat{x}(t) - A\hat{x}(t) - BK\hat{x}(t) + LCe(t) \\ & \dot{e}(t) = [A + LC]e(t) \end{aligned} \tag{34}$$

Considering the augmented state vector defined by: $z(t)\begin{bmatrix} x(t) \\ e(t) \end{bmatrix}$, we can write:

$$\dot{z}(t) = \begin{bmatrix} \dot{x}(t) \\ \dot{e}(t) \end{bmatrix} = \begin{bmatrix} A + BK & -BK \\ 0 & A + LC \end{bmatrix} \begin{bmatrix} x(t) \\ e(t) \end{bmatrix} \tag{35}$$

Consider a quadratic Lyapunov function defined by:

$V(t) = z^T \begin{bmatrix} P & 0 \\ 0 & Q \end{bmatrix} z$, with: P and Q are positive definite matrices.

By deriving this function, we obtain:

$$
\begin{aligned}
\dot{V}(z) &= \dot{z}^T\begin{bmatrix} P & 0 \\ 0 & Q \end{bmatrix} z + z^T\begin{bmatrix} P & 0 \\ 0 & Q \end{bmatrix}\dot{z} \\
&= \begin{bmatrix} x^T & e^T \end{bmatrix}\begin{bmatrix} A^T + K^T B^T & 0 \\ -K^T B^T & A^T + C^T L^T \end{bmatrix}\begin{bmatrix} P & 0 \\ 0 & Q \end{bmatrix}\begin{bmatrix} x \\ e \end{bmatrix} \\
&+ \begin{bmatrix} x^T & e^T \end{bmatrix}\begin{bmatrix} P & 0 \\ 0 & Q \end{bmatrix}\begin{bmatrix} A + BK & -BK \\ 0 & A + LC \end{bmatrix}\begin{bmatrix} x \\ e \end{bmatrix} \\
&= \begin{bmatrix} x^T & e^T \end{bmatrix}\begin{bmatrix} A^T P + K^T B^T P & 0 \\ -K^T B^T P & A^T Q + C^T L^T Q \end{bmatrix}\begin{bmatrix} x \\ e \end{bmatrix} + \\
&+ \begin{bmatrix} x^T & e^T \end{bmatrix}\begin{bmatrix} PA + PBK & -PBK \\ 0 & QA + QCL \end{bmatrix}\begin{bmatrix} x \\ e \end{bmatrix} \\
&= \begin{bmatrix} x^T & e^T \end{bmatrix}\begin{bmatrix} A^T P + K^T B^T P + PA + PBK & -PBK \\ -K^T B^T P & A^T Q + C^T L^T Q + QA + QLC \end{bmatrix}\begin{bmatrix} x \\ e \end{bmatrix}
\end{aligned} \tag{36}
$$

According to the Upper bound theorem, we have:

$$
-e^T K^T B^T Px - x^T PBKe \leq x^T PBB^T Px + e^T K^T Ke \tag{37}
$$

Hence:

$$
\begin{aligned}
\dot{V}(z) &= x^T (A^T P + K^T B^T P + PA + PBK)x - e^T K^T B^T Px \\
&- x^T PBKe + e^T (A^T Q + C^T L^T Q + QA + QLC)e \\
&\leq x^T (A^T P + K^T B^T P + PA + PBK + PBB^T P)x \\
&+ e^T (A^T Q + C^T L^T Q + QA + QLC + K^T K)e
\end{aligned} \tag{38}
$$

So if:

$$
\begin{aligned}
&x^T (A^T P + K^T B^T P + PA + PBK + PBB^T P)x \\
&+e^T (A^T Q + C^T L^T Q + QA + QLC + K^T K)e \prec 0
\end{aligned}
$$

Then $\dot{V}(z) \prec 0$.
By applying the complement of Schur these equations can be written as follows:

$$
\begin{aligned}
&\begin{bmatrix} XA^T + Y^T B^T + Ax + BY & B \\ B^T & -I \end{bmatrix} \prec 0 \\
&\begin{bmatrix} A^T Q + QA + Y_0 C + C^T Y_0^T & K^T \\ K & -I \end{bmatrix} \prec 0
\end{aligned} \tag{39}
$$

The first equation is not linear. By multiplying it on both sides by: $\begin{bmatrix} P^{-1} & 0 \\ 0 & I \end{bmatrix}$, we obtain:

$$\begin{bmatrix} P^{-1} & 0 \\ 0 & I \end{bmatrix}\begin{bmatrix} A^T P + K^T B^T P + PA + PBK & PB \\ B^T P & -I \end{bmatrix}\begin{bmatrix} P^{-1} & 0 \\ 0 & I \end{bmatrix} \prec 0 \Rightarrow$$
$$\begin{bmatrix} P^{-1}A^T P + P^{-1}K^T B^T P + P^{-1}PA + P^{-1}PBK & P^{-1}PB \\ B^T P & -I \end{bmatrix}\begin{bmatrix} P^{-1} & 0 \\ 0 & I \end{bmatrix} \prec 0 \Rightarrow$$
$$\begin{bmatrix} P^{-1}A^T + P^{-1}K^T B^T + AP^{-1} + BKP^{-1} & B \\ B^T & -I \end{bmatrix} \prec 0$$

By posing:

$$X = P^{-1}$$
$$Y = KX$$
$$Y_0 = LQ$$

We obtain:

$$\begin{bmatrix} XA^T + Y^T B^T + Ax + BY & B \\ B^T & -I \end{bmatrix} \prec 0$$
$$\begin{bmatrix} A^T Q + QA + Y_0 C + C^T Y_0^T & K^T \\ K & -I \end{bmatrix} \prec\sim 0$$

The gains are given by:

$$K = YX^{-1}$$
$$L = Q^{-1}Y_0$$

5 Conclusion

The Linear Matrix Inequalities (LMIs) are discussed in this chapter. There was also a discussion of the words "observability" and "observer." A linear observer-based control system employing the LMIs approach is also described.

References

1. Chang, S.S.L., Peng, T.K.C.: Adaptive guaranteed cost control systems with uncertain parameters. IEEE Trans. Autom. Control **17**, 474–483 (1972)
2. Wonham, W.M.: Optimal stationary control of linear system with state dependent noise. SIAM J. Control (1967)
3. Camozzi, P.: Uncertain Modeling Systems: control by Placement. of pole, Thesis, Paul Sabatier University of Toulouse, France (1995)
4. Bien, Z., Kim, J.H.: A robust stability bound of linear system with structured uncertainty. IEEE Trans. Autom. Control **37**(10), 1549–1551 (1992)
5. Gao, Z., Antsaklis, P.J.: Explicit asymmetric bounds for robust stability of continuous and discret-time systems. IEEE Trans. Autom. Control **38**, 332–335 (1993)
6. Lyapunov, A.M.: General problem of stability of movement. Ann. Fac. Sci. Toulouse **9**, 203–474 (1907)
7. Hahn, W.: Stability of Motion. Springer, Berlin (1967)
8. Parks, P.C., Hahn, V.: Stabililtäts Theorie. Springer, Berlin (1981)
9. Khalil, H.K.: Nonlinear System. Prentice Hall, Upper Saddle River, N. J. 07458 (1996)
10. Narendra, K.S., Annaswamy, A.M.: Stable Adaptive Systems. Prentice Hall Information ans System Sciences Series, N. J. 07632 (1989)
11. Borne, P., Dauphin-Tanguy, G., Richard, J.P., Rotella, F., Zambettakis, I.: Analysis and Regulation of Industrial Processes. Technip, Paris (1993)
12. Lien, C.H.: An efficient method to design robust observer-based control of uncertain linear systems. Appl. Math. Comput. **158**, 29–44 (2004)
13. Hua, C., Li, F., Guan, X.: Observer-based adaptive control for uncertain time delay systems. Inf. Sci. (2004). (Article in press)
14. Chadli, M., Maquin, D., Ragot, J.: On the stability of multiple model systems. In: Proceedings of the European Control Conference, pp. 1894–1899 (2001)
15. Tanaka, K., Ikeda, T., He, Y.Y.: Fuzzy regulators and fuzzy observers: relaxed stability conditions and LMI-based design. IEEE Trans. Fuzzy Syst. **6**(1), 250–256 (1998)
16. Luenberger, D.G.: Determining the state of the linear system with observers of low dynamic order, Ph.D. dissertation, Departement of Elec. Engineers., Stanford University, California (1963)
17. Lewis, F.L.: A survey of linear singular systems. Circuits Syst. Signal Process **5**, 3–36 (1986)
18. Chen, J., Zhang, HH.: Robust detection of faulty actuators via unknown input observers. Int. J. Syst. Sci. **22**, 1829–1839 (1991)
19. Soffker, D.: Fault detection using proportional integral observers for application to elastic mechanical structures. In: Proceeding of the Multiconference CESA IMACS, Symposium on Control, Optimization and Supervision. Lille, France, pp. 522–527 (1996)
20. Jiang, G.P., Wang, S.P., Song, W.Z.: Design of observer with intergrators for linear systems with unknown input disturbances. Electron. Lett. **36**(13), 1168–1169 (2000)
21. Chadli, M., Maquin, D., Ragot, J.: Nonquadratic stability analysis of Takagi Sugeno models. IEEE Conf. Decision Control **2**, 2143–2148 (2002)
22. Karamarkar, N.: A new polynomial time algorithm for linear programming. Combinatorica **4**(4), 373–395 (1984)
23. Nesterov, Y., Balakrishnan, J.: A common Lyapunov function for stable LTI systems with commuting A-matrices. IEEE Trans. Autom. Control **39**(12), 2469–2471 (1994)
24. Hassibi, A., How, J., Boyd, S.: A path-following method for solving BMI problems in control. In: American Control Conference, vol. 2, pp. 1385–1389. San Diego, California (1999)
25. Boyd, S., El Ghaoui, L., Feron, E., Balakrishnan, V.: Linear Matrix Inequalities in System and Control Theory. SIAM, Philadelphia (1994)
26. Chilali, M., Gabinet, P.: H∞ Design with pole placement constraints: an LMI approach. IEEE Trans. Autom. Control **41**(3), 358–367 (1996)
27. Fossard, A.J., Normand-Cyrot, D.: Système Non Linéaire, Tome 1 : Modeling-Estimation. Masson, Paris (1993)

28. O'Reilly, J.: Observer for Linear System, Richard Bellman, Mathematics in Science and Engineering, vol. 140. Academic Press, New York (1983)
29. Follinger, O.: Regelungstechnik—Einführung in die methoden und ihre anwendungen. Dr. Alfred Huthig Verlag, Heidelberg (1985)
30. Borne, P., Dauphin-Tanguy, G., Richard, J.P., Rotella, F., Zambettakis, I.: Modeling and Identification of Processes, Tome 1. Editions Technip, Paris (1992)
31. Kalman, R.E., Betram, J.E.: Control system analysis and design via the "second method" of Lyapunov -I : Continuous-time system. ASME J. Basic Eng. **82**, 371–393 (1960)
32. Borne, P., Dauphin-Tanguy, G., Richard, J.P., Rotella, F., Zambettakis, I.: Control and Optimization of Processes, Collection of Engineering Methods and Techniques. Editions Technip, Paris (1990)
33. Magni, J.F., Mouyon, P.: A generalized approach to observers for fault diagnosis. In: Proceedings of the 30th Conference on Decision and Control, vol. 3, pp. 2236–2241. Brignton, England (1991)
34. Dieulessaint, E., Royer, D.: Linear control systems with Brignton sampled signals. Autom. Appl. Masson (1994)
35. Flugge-Lotz, I., Taylor, C.: Synthesis of nonlinear control systems. IRE Trans. Autom. Control **1**(1), 3–9 (1956)
36. Letov, A.M.: Conditionally stable control systems (on a class of optimal control systems). Autom. Remote Control **7**, 649–664 (1957)
37. Maslennikov, V.A.: High-quality control of neutral plants by means of an astatic controller. Autom. Remote Control **2** (1956)
38. Ostrovsky, G.M.: Application of nonlinear correction to control systems of the second order. Autom. Remote Control **11** (1956)
39. Utkin, V.I.: Principles of identification using sliding regimes. Soviet Trans. Autom. Control **26**, 271–272 (1981)
40. Utkin, V.I.: Variable structure systems with sliding modes. IEEE Trans. Autom. Control **22**(22), 212–222 (1977)

Some Preliminaries on Unknown Input Observers, Discontinuous Observers and Sliding Mode Observers Design

Boutheina Maalej, Jalloul Méliani, Omar Naifar, Nabil Derbel, and Abdellatif Ben Makhlouf

Abstract In this chapter, some preliminaries on unknown input observers are given. Furthermore, the development of discontinuous observers and the methods in which they are designed are introduced. The final parts of the chapter are dedicated to the sliding mode observers design.

Keywords Linear system · Observers · Observer design · Sliding mode

1 Introduction

The problem of optimal process control, when some state vector components are not measurable, has undoubtedly initiated the first works on observers. These allow the development of a state estimation model using the accessible variables of the system, such as its inputs and outputs.

In the deterministic case, this model is known as a state observer [1, 2] and in the case of a stochastic system, this model is called a filter [3–5].

B. Maalej · N. Derbel
Laboratory of Control & Energy Management, Digital Research Center of Sfax, University of Sfax, ENIS, Sfax, Tunisia
e-mail: n.derbel@enis.rnu.tn

J. Méliani
Higher Institute of Apllied Sciences and Technology of Kairouan, University of kairouan, kairouan, Tunisia

O. Naifar (✉)
Laboratory of Control & Energy Management, Department of Electrical Engineering, National School of Engineering, University of Sfax, Sfax, Tunisia

A. Ben Makhlouf
Department of Mathematics, College of Science, Jouf University, P.O. Box: 2014, Sakaka, Saudi Arabia
e-mail: abmakhlouf@ju.edu.sa

O. Naifar and A. Ben Makhlouf (eds.), *Advances in Observer Design and Observation for Nonlinear Systems*, Studies in Systems, Decision and Control 410,
https://doi.org/10.1007/978-3-030-92731-8_3

This state estimation uses the measured outputs of the system, its inputs and its model. When a system is completely observable, the state reconstruction can be performed either by a full order observer (the order of the observer is the same as the one of the system), or by a reduced order observer (the order of the observer is smaller than the one of the system).

The asymptotic convergence of the state estimation error to zero requires a very precise determination of the observer matrices. Raymond [6] has shown that a small error on the parameters of the system matrices could generate a large reconstruction, an important reconstruction error (obtained by comparing the estimate to the measured ones). Several authors have presented state estimation techniques based on the design of proportional and integral action observers for uncertain linear systems [7] and singular systems [8, 9]. In the presence of unknown inputs and sensor faults, there are several techniques of state estimation which will be discussed in this chapter.

2 Unknown Input Observer

A physical process is often subject to disturbances which have as origin the noise due to the process environment, the measurement uncertainties, sensor or actuator faults; these disturbances have adverse effects on the normal behavior of the process and these estimation can be used to design a controlled system able to minimize their effects. Disturbances are called unknown inputs when they affect the process input and their presence can make it difficult to estimate the system state.

Several works have been done concerning the estimation of the state and the output in the presence of unknown inputs and they can be grouped into two categories. The first assumes a priori knowledge of information about these unmeasurable inputs, in particular, Johnson [10] has proposed a polynomial approach and Meditch [11] has suggested approximating the unknown inputs by a known dynamic system response. The second category proceeds either by estimating the unknown input [12], or by its complete elimination from the system equations [13, 14].

Among the techniques that do not require the elimination of unknown inputs, several authors have proposed observer design methods capable of fully reconstructing the state of a linear system in the presence of unknown inputs [15, 16]; Kobayashi [17], Lyubchik [18] and Liu [19] have used a model inversion method for state estimation.

Besides, among the techniques that allow the elimination of unknown inputs, the one proposed by Kudva [20] is interested, in the case of linear systems, in the existence conditions of the unknown input system observer based on the technique of the generalized matrix technique. Guan has proceeded to the elimination of the unknown inputs of the state equations for continuous linear systems [21]. Several other variants exist, but the majority of them have been developed for linear systems.

Koenig [8] has presented a simple method to design a proportional and integral action observer for singular systems with unknown inputs. Sufficient conditions for the existence of this observer have been established.

Reduced order observers have been considered by several authors in recent years [22–24]. However, Yang and Wilde [22] have demonstrated that the full order unknown input observer can have a faster convergence speed than the reduced order observer.

The use of unknown input observers for fault diagnosis and process monitoring systems has also attracted a lot of attention [13, 24–26] and [27]. Dassanayake, [13] has considered an observer, by eliminating unknown inputs in the state equations, to be able to detect and isolate several sensor faults, in the presence of unknown inputs, on an engine (turbojet).

2.1 State Reconstruction by Eliminating Unknown Inputs

The reconstruction of the linear dynamical system state where several inputs are not measurable is of a great interest in practice. In such circumstances, a conventional observer, which requires the knowledge of the inputs, cannot be used directly. The Unknown Inputs Observer (UIO) has been developed to estimate the system state, despite the existence of unknown inputs or disturbances by eliminating them in the state equations. This type of observer has attracted the attention of many researchers [10, 16, 19, 28, 29].

In this section, we show that the convergence conditions of an unknown input observer are solutions of bilinear matrix inequalities (BMI) which can be linearized by different techniques to obtain linear matrix inequalities (LMI).

2.2 Reconstruction Principle

Consider the linear dynamic system with unknown inputs, described by the following equations :

$$\begin{cases} \dot{x}(t) = Ax(t) + Bu(t) + R\bar{u}(t) \\ y(t) = Cx(t) \end{cases} \tag{1}$$

where $x(t) \in \mathbb{R}^n$ is the state vector, $u(t) \in \mathbb{R}^m$ is the vector of known inputs, $\bar{u}(t) \in \mathbb{R}^q$, $q < n$ is the vector of unknown inputs, $y(t) \in \mathbb{R}^p$ represents the vector of measurable outputs. $A \in \mathbb{R}^{n\times n}$ is the state matrix of the linear system, $B \in \mathbb{R}^{n\times m}$ is the input matrix, $R \in \mathbb{R}^{n\times q}$ is the influence matrix of the unknown inputs and $C \in \mathbb{R}^{p\times n}$ is the output matrix.

We assume that the matrix R is of full column rank and that the pair (A, C) is observable. The objective is the complete estimation of the state vector despite the presence of the unknown inputs $\bar{u}(t)$. Thus, consider the full order observer [30] :

$$\begin{cases} \dot{z}(t) = Nz(t) + Gu(t) + Ly(t) \\ \hat{x}(t) = z(t) - Ey(t) \end{cases} \tag{2}$$

where $z(t) \in \mathbb{R}^n$ is the state vector, $\hat{x}(t) \in \mathbb{R}^n$ is the estimate of the state vector $x(t)$. In order to guarantee this estimation, $\hat{x}(t)$ must asymptotically approach to $x(t)$, that is the state estimation error

$$e(t) = x(t) - \hat{x}(t) \tag{3}$$

approaches to zero asymptotically. The dynamics equation of the evolution of this error is written as follows:

$$\begin{aligned} \dot{e}(t) &= \dot{x}(t) - \dot{z}(t) + EC\dot{x}(t) \\ &= (I + EC)\dot{x}(t) - \dot{z}(t) \\ &= (I + EC)(Ax(t) + Bu(t) + R\bar{u}(t)) - (Nz(t) + Gu(t) + Ly(t)) \\ &= (I + EC)(Ax(t) + Bu(t) + R\bar{u}(t)) - (N\hat{x}(t) + Gu(t) + (L + NE)Cx(t)) \end{aligned} \tag{4}$$

Let us consider $P = I + EC$, then we obtain :

$$\dot{e}(t) = Ne(t) + (PB - G)u(t) + PR\bar{u}(t)) + (PA - NP - LC)x(t) \tag{5}$$

The state estimation error converges asymptotically to zero if and only if:

$$LC = PA - NP \tag{6a}$$

$$G = PB \tag{6b}$$

$$PR = 0 \tag{6c}$$

$$N \ is \ stable^1 \tag{6d}$$

The numerical solution of the system of equation (6) is based on the computation of the pseudo-inverse of the (CR) matrix, this is possible if the matrix (CR) is of full row rank [31].

$$E = -R(CR)^T((CR)(CR)^T)^{-1} \tag{7a}$$

$$P = I - R(CR)^T((CR)(CR)^T)^{-1}C \tag{7b}$$

$$G = PB \tag{7c}$$

$$N = PA - KC \tag{7d}$$

$$L = K - NE \tag{7e}$$

$$N \ is \ stable \tag{7f}$$

Thus, if the system of equation (7) is satisfied, the dynamics of the state estimation error reduces to :

$$\dot{e}(t) = Ne(t) \tag{8}$$

Given the properties of N, the state estimation error converges well asymptotically to zero.

2.3 Convergence Conditions of the Observer

In this section, we develop sufficient conditions for the asymptotic convergence of the state estimation error to zero. According to (8), this convergence is guaranteed if there exists a symmetric and positive definite matrix X, such that

$$N^T X + XN < 0 \tag{9}$$

Since $N = PA - KC$, the inequality (9) becomes :

$$(PA - KC)^T X + X(PA - KC) < 0 \tag{10}$$

Unfortunately, we notice that the previous inequality (10) has the disadvantage of being non-linear (bilinear) with respect to the variables K and X. Two methods of resolution can be used:

- Linearization with respect to the variables K and X,
- Change of variables.

2.4 Resolution Methods

Solving methods have been proposed to solve nonlinear matrix inequalities and in particular the bilinear ones [32].

2.5 Linearization with Respect to Variables

We can use a "local" method, based on the linearization of the inequalities, with respect to the variables K and X, around the initial values K_0 and X_0 (well chosen). We define:

$$K = K_0 + \partial K \quad \text{and} \quad X = X_0 + \partial X \tag{11}$$

From the inequality (10), we obtain :

$$\begin{cases} ((PA-(K_0+\Delta K)C)+(PA-(K_0+\Delta K)C^T)(X_0+\Delta X)+ \\ (X_0+\Delta X)((PA-(K_0+\Delta K)C)+(PA-(K_0+\Delta K)C^T))<0 \\ X_0+\Delta X>0 \end{cases} \tag{12}$$

Ignoring the second order terms of the inequality (12), we obtain:

$$\begin{cases} ((PA-K_0C+(PA-K_0C)^T)\Delta X+\Delta X((PA-K_0C)+(PA-K_0C)^T)- \\ \Delta KCX_0-(CX_0)^T\Delta K^T-C^T\Delta K^TX_0-X_0\Delta KC+ \\ ((PA-K_0C)+(PA-K_0C)^T)X_0+X_0((PA-K_0C)+(PA-K_0C)^T)<0 \\ X_0+\Delta X>0 \end{cases} \tag{13}$$

The system (13) is then a LMI (linear matrix inequality) type problem and its solution with respect to ΔK and ΔX is standard [33]. Note that the choice of initial values K_0 and X_0 remains the main drawback of this method and moreover the convergence to a solution is not always guaranteed. Unfortunately, from a practical point of view, one may have to examine various choices of initial values in order to to obtain a solution.

Remark 1 The LMI system (13) is valid only in the neighborhood of K_0 and X_0; this encouraged us, in order to improve the resolution, to propose, to limit the variations of the matrices δK and δX, the following additional constraints:

$$\begin{cases} \|\Delta K_0\| < \epsilon\|K_0\|, \\ \|\Delta X_0\| < \epsilon\|X_0\| \quad with \quad 0<\epsilon \ll 1. \end{cases} \tag{14}$$

The LMI formulation of these constraints (14) is described by the following matrix inequalities:

$$\begin{bmatrix} \epsilon\|X_0\|I_{n\times n} & \Delta X \\ \Delta X & \epsilon\|X_0\|I_{n\times n} \end{bmatrix} > 0, \quad \begin{bmatrix} \epsilon\|X_0\|I_{n\times n} & \Delta K \\ \Delta K & \epsilon\|X_0\|I_{m\times m} \end{bmatrix} > 0. \tag{15}$$

If the LMI systems (13) and (15) are feasible, then the observer (2) asymptotically estimates the state of the linear system with unknown inputs (1).

2.5.1 Change of Variables

To overcome the drawbacks of the previous method, a method based on a variable change is more interesting. For that, let us consider the following change of variables:

$$W = XK \tag{16}$$

The inequality obtained after this change of variables can be written as follows:

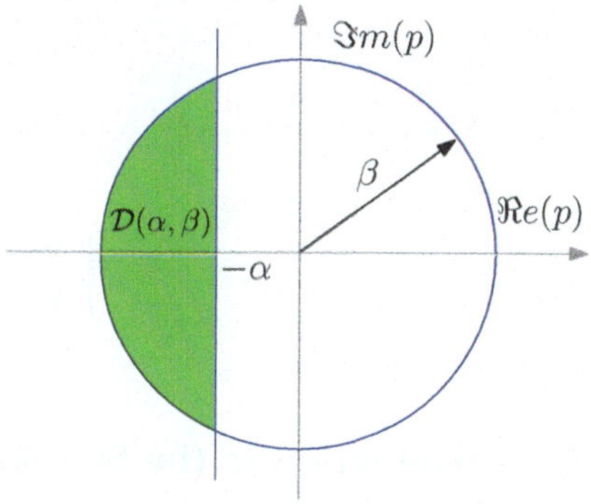

Fig. 1 LMI area

$$(PA)^T X + X(PA) - (C^T W^T + WC) < 0 \tag{17}$$

The solution of the initial problem is obtained in two steps. First, we solve the linear matrix inequality (17) with respect to the unknowns X and W. Then we deduce the value of the gain K by the formula :

$$K = X^{-1} W \tag{18}$$

2.6 *Pole Placement*

In this section, we examine how to improve the performance of the observer in particular with respect to the convergence speed to zero of the state estimation error.

For a better estimation of the state, the observer dynamics is chosen to be faster than that of the system. For this, we fix the eigenvalues of the observer in the left half-plane of the complex plane so that their real parts are larger in absolute value than those of the state matrix.

To ensure some convergence dynamics of the state estimation error, we define the complex region $\mathcal{D}(\alpha, \beta)$ by the intersection of a circle with center (0, 0) and radius equal to β and the left half of the region bounded by a vertical line of coordinates equal to $-\alpha$ where α is a positive constant (Fig. 1).

2.6.1 Corollary

The eigenvalues of the matrix N are in the LMI region $\mathcal{D}(\alpha, \beta)$ if there exist matrices ΔX and ΔK such that:

$$\begin{bmatrix} -\beta(X_0 + \Delta X) & N_0^T X - (\Delta K C)^T X_0 \\ X N_0 - X_0(\Delta K C) & -\beta(X_0 + \Delta X) \end{bmatrix} < 0 \quad (19)$$

$$N_0^T \Delta X + \Delta X N_0 - C^T \Delta K^T X_0 - X_0 \Delta K C + N_0^T X_0 + X_0 N_0 + 2\alpha(X_0 + \Delta X) < 0$$

with

$$\begin{cases} N_0 = PA - K_0 C, \\ X = X_0 + \Delta X \end{cases} \quad (20)$$

3 Introduction to the Development of Discontinuous Observers

In recent years, the control problem or diagnosis of uncertain dynamic systems subject to external disturbances has been the subject of great interest. In practice, it is not always possible to measure the state vector, in this case, a design method based only on measured outputs and known inputs is used.

From a robust control perspective, the desirable properties of variable structure control systems, especially with a sliding mode, are well developed [34, 35]. Despite the successful research and development activity of variable structure control theory and its insensitivity to uncertainties or unknown inputs, few authors have considered the application of the fundamental principles to the observer design problem. Utkin has presented the design of an observer method with a discontinuous structure for which the error between the estimated and measured outputs is forced to converge to zero [36]. Dorling and Zinober [37] have explored the practical application of this observer to an uncertain system and examine the difficulties of choosing an appropriate sliding gain. Walcott et al. [38], Walcott and Zac [39] and Zak [40] have presented a method of observer design based on the Lyapunov approach. Under appropriate assumptions, they have shown the asymptotic decay of the state estimation error in the presence of bounded nonlinearities/uncertainties.

Recently, Ha [41] has presented a methodology to design a sliding mode controller for an uncertain linear system based on the pole placement technique. Xiong [42] has considered a sliding mode observer for the state estimation of an uncertain nonlinear system, the uncertainties are considered as unknown inputs. Islam [43] has proposed a theoretical and experimental evaluation of a sliding mode observer to measure the position and the velocity on a switched reluctance motor. In this section, we seek to construct a sliding mode observer building on the existing contributions described above. A detailed reminder of the design approaches of Utkin [34] and Walcott and Zac [39, 44] has been provided. Then, we are interested in the methodology developed by Edwards and Spurgeon [45, 46] to determine the gain expression of an observer, which overcomes the drawbacks of the observer of Walcott and Zak [39]. A Lyapunov approach have been proposed to ensure asymptotic convergence of the state estimation error. The solution of the Lyapunov inequalities leads to the solution of a LMI type problem.

4 Methods for Discontinuous Observer Design

Considering the following uncertain dynamic system:

$$\begin{cases} \dot{x}(t) = Ax(t) + Bu(t) + f(x, u, t) \\ y(t) = Cx(t) \end{cases} \tag{21}$$

where $x(t)$ is the state vector, $u(t)$ is the vector of known inputs, $y(t)$ represents the measurable output. $A \in \mathbb{R}^{n\times n}$, $B \in \mathbb{R}^{n\times m}$, $C \in \mathbb{R}^{p\times n}$ with $p \geq m$. The unknown function $f : \mathbb{R}^n \times \mathbb{R}^m \times \mathbb{R}_+ \to \mathbb{R}^n$ represents the uncertainties and satisfies the following conditions.

$$||f(x, y, t)|| \leq \rho, \forall x(t) \in \mathbb{R}^n, \quad u(t) \in \mathbb{R}^m, \quad t \geq 0 \tag{22}$$

Moreover, the matrix C is assumed to have full row rank. The problem considered here is the reconstruction of the state vector in spite of the presence of unknown inputs.

4.1 Utkin Observer

Consider first the system (22) and assume that the pair (A, C) is observable and that the function $f(x, u, t) \equiv 0$. Since the state reconstruction relies on measured outputs, it is natural to perform a coordinate change so that the system outputs appear directly as components of the state vector. Without loss of of generality, the output matrix can be written as follows:

$$C = \begin{bmatrix} C_1 \; C_2 \end{bmatrix} \tag{23}$$

where $C_1 \in \mathbb{R}^{p\times(n-p)}$, $\quad C_2 \in \mathbb{R}^{p\times p}$, $\quad withdet(C_2) \neq 0$, then the transformation matrix

$$T^{-1} = \begin{bmatrix} I_{n-p} & 0 \\ -C_1C_2^{-1} & C_2^{-1} \end{bmatrix} \tag{24}$$

is non-singular and, in this new coordinate system, we can easily verify that the new output matrix is written as follows:

$$CT^{-1} = \begin{bmatrix} 0 \; I_p \end{bmatrix} \tag{25}$$

The new state and control matrices are expressed as:

$$A = TAT^{-1} = \begin{bmatrix} A_{11} & A_{12} \\ A_{21} & A_{22} \end{bmatrix} \quad and \quad B = TB = \begin{bmatrix} B_1 \\ B_2 \end{bmatrix} \tag{26}$$

The nominal system can then be written as follows:

$$\begin{cases} \dot{x_1}(t) = A_{11}x_1(t) + A_{12}y(t) + B_1u(t) \\ \dot{y}(t) = A_{21}x_1(t) + A_{22}y(t) + B_2u(t) \end{cases} \tag{27}$$

where

$$\begin{bmatrix} x_1(t) \\ y(t) \end{bmatrix} = Tx(t) \quad and \quad x_1(t) \in \mathbb{R}^{n-p}. \tag{28}$$

The observer proposed by Utkin [36] has the following form:

$$\begin{cases} \dot{\hat{x}}_1(t) = A_{11}\hat{x}_1(t) + A_{12}\hat{y}(t) + B_1u(t) + L\upsilon(t) \\ \dot{\hat{y}}(t) = A_{21}\hat{x}_1(t) + A_{22}\hat{y}(t) + B_2u(t) - \upsilon(t) \end{cases} \tag{29}$$

where $(\hat{x}_1(t), \hat{y}(t))$ are the estimated values of $(x_1(t), y(t))$, $L \in \mathbb{R}^{(n-p)\times p}$ is the gain observer and the components of the discontinuous vector $\upsilon(t)$ are defined by the following equation :

$$\upsilon_i(t) = Msign(\hat{y}_i(t) - y_i(t)), \quad for \;\; M \;\; \in \;\; \mathbb{R}_+ \tag{30}$$

where $\hat{y}_i(t)$ and $y_i(t)$ are the components of the vectors $\hat{y}(t)$ and $y(t)$ respectively and $sign$ is the signum function.

Let us denote by $e_1(t)$ and $e_y(t)$ the state and output estimation errors.

$$\begin{aligned} e_1(t) &= \hat{x}_1(t) - x_1(t) \\ e_y(t) &= \hat{y}(t) - y(t) \end{aligned} \tag{31}$$

From Eqs. (27), (29) and (31), the following system can be obtained:

$$\begin{cases} \dot{e}_1(t) = A_{11}e_1(t) + A_{12}e_y(t) + L\upsilon(t) \\ \dot{e}_y(t) = A_{21}e_1(t) + A_{22}e_y(t) + \upsilon(t) \end{cases} \tag{32}$$

As the pair (A, C) is observable, so is the pair (A_{11}, A_{21}). Therefore, L can be chosen so that the eigenvalues of the matrix $A_{11} + LA_{21}$ are in the left half-plane of the complex plane. Now let us define the new change of variable:

$$T_s^{-1} = \begin{bmatrix} I_{n-p} & -L \\ 0 & I_p \end{bmatrix} \; with \; \begin{bmatrix} x_1'(t) \\ y(t) \end{bmatrix} = T_s \begin{bmatrix} x_1(t) \\ y(t) \end{bmatrix} \tag{33}$$

After this change of variable, the estimation errors can be written as:

$$\dot{e}_1'(t) = A_{11}'e_1'(t) - A_{12}'e_y(t) \tag{34}$$

$$\dot{e}_y'(t) = A_{11}'e_1'(t) - A_{22}'e_y(t) - \upsilon(t) \tag{35}$$

with $\dot{e}_1'(t) = e_1(t) - Le_y(t)$ and $A_{11}' = A_{11} + LA_{21}$, $A_{12}' = A_{12} + LA_{22} - A_{11}'L$ and $A_{22}' = A_{22} - A_{21}L$. It can be shown, using the theory of singular perturbations, that for a M large enough, a sliding motion can arise on the output error (35). Thus, after a finite time t_s, the error $e_y(t)$ and its derivative are zero ($e_y(t) = 0$, $\dot{e}_y(t) = 0$). Equation (34) becomes:

$$\dot{e}_1'(t) = A_{11}'e_1'(t) \tag{36}$$

By correctly choosing the gain matrix L (so that the matrix A_{11}' is stable), the system of error Eqs. (34)–(35) is stable, i.e., $e_1'(t) \to 0$ when $t \to \infty$.

Therefore $\hat{x}_1(t) \to x_1(t)$ and the other component of the state vector $x_2(t)$ can be reconstructed in the original coordinate system as follows:

$$\hat{x}_2(t) = C_2^{-1}(y(t) - C_1\hat{x}_1(t)) \tag{37}$$

The main practical difficulty of this approach lies in the choice of an appropriate gain M to induce a sliding motion in a finite time. Dorling and Zinober [37] have shown the need to modify the gain M during the time interval in order to reduce excessive switching.

4.2 Walcott and Żak Observer

The problem considered by Walcott and Żak [39, 44] is the state estimation of a system described by (21) such that the error goes to zero exponentially despite the presence of the considered uncertainties. In this part, we assume that :

$$f(x, u, t) = R\xi(x, t) \tag{38}$$

where $\xi : \mathbb{R}^n \times \mathbb{R}_+ \to \mathbb{R}^q$ is a bounded and unknown function, such that :

$$\|\xi(x(t), t)\| \leq \rho, \quad \forall x(t) \in \mathbb{R}_n, \quad t \geq 0$$

Consider a matrix $G \in \mathbb{R}^{n \times p}$ such that the matrix $A_0 = (A - GC)$ has stable eigenvalues, a pair of symmetric, positive and definite Lyapunov matrices (P, Q) and a matrix F respecting the following structural constraint:

$$\begin{aligned} (A - GC)^T P + P(A - GC) &= -Q \\ C^T F^T &= PR \end{aligned} \tag{39}$$

The proposed observer can be expressed as:

$$\dot{\hat{x}} = A\hat{x}(t) + Bu(t) - G(C\hat{x}(t) - y(t)) + \upsilon(t) \tag{40}$$

$$\upsilon(t) = \begin{cases} -\rho \dfrac{P^{-1}C^T F^T FCe(t)}{\|FCe(t)\|} & \text{if } FCe(t) \neq 0 \\ 0 \text{ otherwise} \end{cases} \tag{41}$$

where

$$e(t) = \hat{x}(t) - x(t) \tag{42}$$

The dynamics of the state estimation error generated by this observer is determined by the following equation:

$$\begin{aligned} \dot{e}(t) &= \dot{\hat{x}}(t) - \dot{x}(t) \\ &= A\hat{x}(t) + Bu(t) - G(C\hat{x}(t) - y(t)) + \upsilon(t) - (Ax(t) + Bu(t) + R\xi(x,t)) \\ &= (A - GC)e(t) + \upsilon(t) - R\xi(x,t) \end{aligned} \tag{43}$$

The following Lyapunov function is considered:

$$V(e)(t) = e^T(t)Pe(t) \tag{44}$$

Its derivative along the trajectory of the estimation error can be written as:

$$\begin{aligned} \dot{V}(e(t)) &= \dot{e}^T(t)Pe(t) + e^T(t)P\dot{e}(t) \\ &= ((A - GC)e(t) + \upsilon(t) - R\xi(x,t))^T Pe(t) + e^T(t)P((A - GC)e(t) + \upsilon(t) - R\xi(x,t)) \\ &= -e^T(t)Qe(t) + 2e^T(t)P\upsilon(t) - 2e^T(t)PR\xi(x,t) \\ &= -e^T(t)Qe(t) + 2e^T(t)P\upsilon(t) - 2e^T(t)C^T F^T \xi(x,t) \end{aligned} \tag{45}$$

Let us consider the two following cases:

First case

If $FCe(t) \neq 0$, by replacing the expression of $\xi(t)$ by Eq. (41), the derivative of the Lyapunov function becomes :

$$\begin{aligned} \dot{V}(e(t)) &= e^T(t)Qe(t) - 2e^T(t)\rho \frac{C^T F^T FCe(t)}{\|FCe(t)\|} - 2e^T C^T F^T \xi(x,t) \\ &= -e^T Qe(t) - 2\rho\|FCe(t)\| - 2e^T(t)C^T F^T \xi(x,t) \end{aligned} \tag{46}$$

Using the fact that the unknown function $\upsilon(x,t)$ is bounded by a positive scalar ρ, the derivative of the Lyapunov function can be increased as follows:

$$\begin{aligned}\dot{V}(e(t)) &\leq -e^T(t)Qe(t) - 2\rho\|FCe(t)\| + 2\rho\|FCe(t)\| \\ &\leq -e^T Qe(t) < 0\end{aligned} \tag{47}$$

Second case If $FCe(t) = 0$, by replacing the expression of $\xi(t)$ by Eq. (41), the derivative of the Lyapunov function becomes :

$$\dot{V}(e(t)) = -e^T(t)Qe(t) < 0 \tag{48}$$

Thus, in both cases, we have shown that the derivative of the Lyapunov function is negative which shows that the state estimation error converges asymptotically to zero. To guarantee the asymptotic convergence of the observer, we must verify that:

- the pair (A, C) is observable,
- there exists a pair of Lyapunov matrices (P, Q) and a matrix F respecting the constraints (39).

5 Sliding Mode Observer Using a Canonical Form

Edward and Spurgeon [45, 46] have presented a method for designing a sliding mode observer, based on the structure of the Walcott and Zak observer [39], while avoiding the major drawback of the Walcott and Zak observer mentioned above. For this purpose, let us consider again the dynamical system presented previously:

$$\begin{cases}\dot{x}(t) = Ax(t) + Bu(t) + R\xi(x, u, t) \\ y(t) = Cx(t)\end{cases} \tag{49}$$

where $A \in \mathbb{R}^{n\times n}$, $B \in \mathbb{R}^{n\times m}$, $C \in \mathbb{R}^{p\times n}$ and $D \in n \times q$ with $p \geq q$. We suppose that the matrices A, B and R are of full rank and the function $\xi : \mathbb{R}_+ \times \mathbb{R}^n \times \mathbb{R}^m \rightarrow \mathbb{R}^q$ is unknown bounded function such that :

$$\|\xi(x, u, t)\| \leq \rho \tag{50}$$

Before proceeding to the estimation of the state and output vector of the system (49), we will proceed to two coordinates changes of the state vectors.

5.1 Simplified Output Equation

Suppose that the system described above is observable. It is quite natural to perform a change of coordinates so that the outputs of the system appear directly as components of the state vector. Without loss of generality, the output matrix can be written as [39]:

$$C = \begin{bmatrix} C_1 \; C_2 \end{bmatrix} \tag{51}$$

where $C_1 \in R^p \times (n-p),\;\; C_2 \in R^{p\times p} \;\; and \;\; det(C_2) \neq 0$.

Let us then perform the following change of coordinates:

$$\tilde{x}(t) = \tilde{T}x(t) \tag{52}$$

where $\tilde{T}$ is a non-singular matrix definite as:

$$\tilde{T} = \begin{bmatrix} I_{np} & 0 \\ C_1 & C_2 \end{bmatrix} \tag{53}$$

In this new coordinate system, we can easily verify that the new output matrix is written :

$$\tilde{C} = C\tilde{T}^{-1} = [0 \;\;\; I_p] \tag{54}$$

Other matrices are transformed as follows:

$$\tilde{A} = \tilde{T}A\tilde{T}^{-1} = \begin{bmatrix} \tilde{A}_{11} & \tilde{A}_{12} \\ \tilde{A}_{21} & \tilde{A}_{22} \end{bmatrix}, \;\; \tilde{B} = \tilde{T}B = \begin{bmatrix} \tilde{B}_1 \\ \tilde{B}_2 \end{bmatrix}, \;\; and \;\; \tilde{R} = \tilde{T}R = \begin{bmatrix} \tilde{R}_1 \\ \tilde{R}_2 \end{bmatrix} \tag{55}$$

The system (49) can then be written as :

$$\begin{cases} \dot{\tilde{x}}(t) = \tilde{A}x(t) + \tilde{B}u(t) + \tilde{R}\xi(x,u,t) \\ y(t) = \tilde{C}\tilde{x}(t) = \tilde{x}_2(t) \end{cases} \tag{56}$$

The change of coordinates allows to express directly the output vector as a function of a part of the state vector.

Then, the constraints (39) and the Lyapunov matrices (P, Q) can be expressed as:

$$\begin{aligned} (\tilde{A} - \tilde{G}\tilde{C})^T\tilde{P} + \tilde{P}(\tilde{A} - \tilde{G}\tilde{C}) = -\tilde{Q} \\ -\tilde{C}^T\tilde{F}^T = \tilde{P}\tilde{R} \end{aligned} \tag{57}$$

$$\begin{cases} \tilde{P} = (\tilde{T}^{-1})^T P\tilde{T}^{-1} \\ \tilde{Q} = (\tilde{T}^{-1})^T Q\tilde{T}^{-1} \\ \tilde{G} = \tilde{T}^{-1}G \end{cases} \tag{58}$$

5.2 Decoupling of the Unknown Function

We can now use a result established by Walcott and Zak regarding the design of a robust observer with respect to the presence of unknown inputs or model uncertainties.

Let the linear model $(\tilde{A}, \tilde{B}, \tilde{R}, \tilde{C})$ be defined by the state Eq. (56) where $\tilde{A}$ is a stable matrix, and let $(\bar{A}, \bar{B}, \bar{R}, \bar{C})$ be related to $(\tilde{A}, \tilde{B}, \tilde{R}, \tilde{C})$ by the following coordinate transformation:

$$\bar{x}(t) = \bar{T}\tilde{x}(t) \tag{59}$$

Matrices $(\bar{A}, \bar{B}, \bar{R}, \bar{C})$ are expressed as:

$$\begin{cases} \bar{A} = \bar{T}\tilde{A}\bar{T}^{-1} = \begin{bmatrix} \bar{A}_{11} & \bar{A}_{12} \\ \bar{A}_{21} & \bar{A}_{22} \end{bmatrix}, & \bar{B} = \bar{T}\tilde{B} = \begin{bmatrix} \bar{B}_1 \\ \bar{B}_2 \end{bmatrix} \\ \bar{R} = \bar{T}\bar{R} = \begin{bmatrix} \bar{R}_1 \\ \bar{R}_2 \end{bmatrix}, & \bar{C} = \tilde{C}\bar{T}^{-1} = \begin{bmatrix} 0 & I_p \end{bmatrix} \end{cases} \tag{60}$$

The constraints (57) and the matrices $(\tilde{P}, \tilde{Q})$ become:

$$\begin{aligned} (\bar{A} - \bar{G}\bar{C})^T\bar{P} + \bar{P}(\bar{A} - \bar{G}\bar{C}) &= -\bar{Q} \\ \bar{C}^T\bar{F}^T &= \bar{P}\bar{R} \end{aligned} \tag{61}$$

$$\begin{cases} \bar{P} = (\bar{T}^{-1})^T\tilde{P}\bar{T}^{-1} \\ \bar{Q} = (\bar{T}^{-1})^T\tilde{Q}\bar{T}^{-1} \\ \bar{G} = \bar{T}^{-1}\tilde{G} \end{cases} \tag{62}$$

First Proposition:
Let consider the linear model $(\tilde{A}, \tilde{B}, \tilde{R}, \tilde{C})$ defined by the state equation (56) for which there exists a pair of matrices $(\tilde{P}, F)$ defined by constraints (39) and (58), then there exists a nonsingular transformation $\bar{T}$ such that the new coordinates of matrices $(\bar{A}, \bar{B}, \bar{R}, \bar{C})$, $(\bar{P}, F)$ have the following properties:

1. $\bar{A} = \begin{bmatrix} \bar{A}_{11} & \bar{A}_{12} \\ \bar{A}_{21} & \bar{A}_{22} \end{bmatrix}$ where $\bar{A}_{11} \in \mathbb{R}^{(n-p)\times(n-p)}$ is a stable matrix
2. $\bar{R} = \begin{bmatrix} 0 \\ P_{22}^* F^T \end{bmatrix}$ where $P_{22} \in \mathbb{R}^{p\times p}$
3. $\bar{C} = \begin{bmatrix} 0 & I_p \end{bmatrix}$
4. The Lyapunov matrix has a block-diagonal structure $\bar{P} = \begin{bmatrix} \bar{P}_1 & 0 \\ 0 & \bar{P}_2 \end{bmatrix}$ with $\bar{P}_1 \in \mathbb{R}^{(n-p)\times(n-p)}$ and $\bar{P}_2 \in \mathbb{R}^{p\times p}$

Proof let the pair $(\tilde{P}, F)$ associated with the linear model $(\tilde{A}, \tilde{B}, \tilde{R}, \tilde{C})$ and let the Lyapunov matrix $\tilde{P}$ written in the following form:

$$\tilde{P} = \begin{bmatrix} \tilde{P}_{11} & \tilde{P}_{12} \\ \tilde{P}_{21} & \tilde{P}_{22} \end{bmatrix} \quad \text{where} \begin{cases} \tilde{P}_{11} \in \mathbb{R}^{(n-p)\times(n-p)} \\ \tilde{P}_{12} \in \mathbb{R}^(n-p) \times p \ \text{ and } \ \tilde{P}_{22} \in \mathbb{R}^p \times p \end{cases} \tag{63}$$

The coordinate change uses the following transformation matrix $\bar{T}$:

$$\bar{T} = \begin{bmatrix} \tilde{P}_{11} & \tilde{P}_{12} \\ 0 & I_p \end{bmatrix} \tag{64}$$

which is nonsingular, the matrix $\tilde{P}_{11}$ being a positive definite symmetric matrix $\tilde{P}_{11} = \tilde{P}_{11}^T > 0$. In the new coordinates, we obtain: $\bar{C} = \tilde{C}\bar{T}^{-1} = \begin{bmatrix} 0 & I_p \end{bmatrix}$. Thus, property 3 is satisfied. From Eq. (58) we obtain: $\tilde{R} = \tilde{P}^{-1}\tilde{C}^{-1}F^T$. If we note:

$$\tilde{P}^{-1} = \begin{bmatrix} P_{11}^* & P_{12}^* \\ P_{21}^* & P_{22}^* \end{bmatrix} \tag{65}$$

we obtain

$$\bar{R} = \bar{T}\tilde{R} = \begin{bmatrix} \tilde{P}_{11} & \tilde{P}_{12} \\ 0 & I_p \end{bmatrix} \begin{bmatrix} \tilde{P}_{11}^* & \tilde{P}_{12}^* \\ \tilde{P}_{21}^* & \tilde{P}_{22}^* \end{bmatrix} \begin{bmatrix} 0 \\ I_p \end{bmatrix} F^T = \begin{bmatrix} 0 \\ \tilde{P}_{22}^* F^T \end{bmatrix} = \begin{bmatrix} 0 \\ \bar{R}_2 \end{bmatrix} \tag{66}$$

Thus, the second property explaining the decoupling of unknown inputs (uncertain function) is proved. If there exists a Lyapunov matrix $\tilde{P}$ that satisfies the constraints (58), then the matrix $\bar{P} = (\bar{T}^{-1})^T \tilde{P}\bar{T}^{-1}$ represents the Lyapunov matrix for the state matrix $\bar{A}_0 = \bar{A} - \bar{G}\bar{C}$ and satisfies the constraint $\bar{C}^T F^T = \bar{P}\bar{R}$. Using a direct calculation, one can easily find :

$$\bar{P} = \begin{bmatrix} \tilde{P}_{11}^{-1} & 0 \\ -\tilde{P}_{12}^T\tilde{P}_{11}^{-1} & I_p \end{bmatrix} \begin{bmatrix} \tilde{P}_{11} & \tilde{P}_{12} \\ \tilde{P}_{12}^T & \tilde{P}_{22} \end{bmatrix} \begin{bmatrix} \tilde{P}_{11}^{-1} & -\tilde{P}_{11}^{-1}\tilde{P}_{12} \\ 0 & I_p \end{bmatrix} = \begin{bmatrix} \tilde{P}_{11}^{-1} & 0 \\ 0 & \bar{P}_2 \end{bmatrix} \tag{67}$$

where $\bar{P}_2 = -\tilde{P}_{12}^T\tilde{P}_{11}^{-1}\tilde{P}_{12}$. Thus, the matrix $\bar{P}$ has the block-diagonal structure shown in property 4. Finally, replacing the matrix $\bar{P}$ (67) in the constraint (61), we obtain:

$$\begin{bmatrix} \bar{A}_{011} & \bar{A}_{012} \\ \bar{A}_{021} & \bar{A}_{022} \end{bmatrix}^T \begin{bmatrix} \bar{P}_1 & 0 \\ 0 & \bar{P}_2 \end{bmatrix} + \begin{bmatrix} \bar{P}_1 & 0 \\ 0 & \bar{P}_2 \end{bmatrix} = \begin{bmatrix} \bar{A}_{011} & \bar{A}_{012} \\ \bar{A}_{021} & \bar{A}_{022} \end{bmatrix} \tag{68}$$

Then

$$\begin{cases} \bar{A}_{011}^T\bar{P}_1 + \bar{P}_1\bar{A}_{011} < 0 \\ \bar{A}_{022}^T\bar{P}_2 + \bar{P}_2\bar{A}_{022} < 0 \end{cases} \tag{69}$$

with

$$\bar{A}_0 = \begin{bmatrix} \bar{A}_{011} & \bar{A}_{012} \\ \bar{A}_{021} & \bar{A}_{022} \end{bmatrix}. \tag{70}$$

$\bar{A}_{011} = \bar{A}_{11} - \bar{G}\bar{C}_{011} = \bar{A}_{11}$, because $(\bar{G}\bar{C})_{11} = 0 \; \forall G \in \mathbb{R}^{n \times p}$ because $\bar{C} = \begin{bmatrix} 0 & I_p \end{bmatrix}$ and therefore the matrix $\bar{A}_{11}$ is stable. Thus property 1 is proved.

6 Sliding Mode Observer

The implementation of control laws based on the nonlinear model of the system, requires the knowledge of the complete state vector of the system at each instant. However, in most cases, only one part of the state is accessible using of sensors.

To reconstitute the complete system state, the idea is based on the use of a software sensor, called observer.

An observer is a dynamic system which from the system input $u(t)$ (the control), the measured output $y(t)$, as well as a priori knowledge of the model, will provide an estimated output state $\hat{x}(t)$ which should tend towards the real state $x(t)$.

One of the best known classes of robust observers is the sliding mode observers [47].

6.1 Design of Sliding Mode Observer

The principle of sliding mode observers consists in remaining the system dynamics with order n using discontinuous functions, to converge to a variety s of dimension $(n - p)$ called sliding surface (p is the dimension of the measurement vector) [47].

The attractiveness of this surface is ensured by sliding conditions. If these conditions are verified, the system converges towards the sliding surface and y moves according to a $(n - p)$ order dynamics.

In the case of sliding mode observers, the dynamics concerned are those of the observation errors $e(t) = x(t) - \hat{x}(t)$.

From their initial values $e(0)$, these errors converge to the equilibrium values in two steps:

- The first step, the observation error trajectory evolves towards the sliding surface on which the errors between the observer output and the real system output (measurements) $e_y = y - \hat{y}$ are equal to zero. This step, which is generally very dynamic, is called the attainment mode.
- In the second step, the observation error trajectory remains on the sliding surface with imposed dynamics, to cancel all the observation errors. This last mode is called sliding mode.

Consider the following n-order nonlinear state system :

$$\begin{cases} \dot{x}(t) = f(x, u) \\ y(t) = h(x) \end{cases} \tag{71}$$

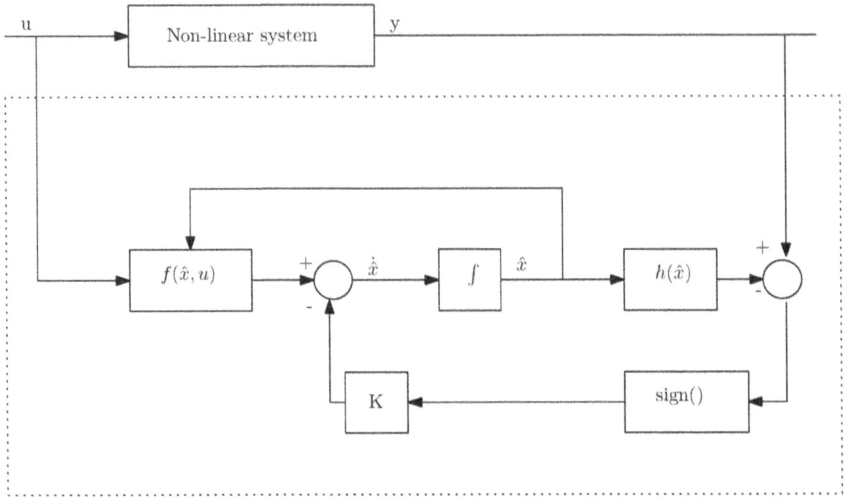

Fig. 2 Block diagram of a sliding mode observer

where $x \in \mathbb{R}^n$ is the state vector, $u \in \mathbb{R}^m$ is the vector of known inputs or control, $y \in \mathbb{R}^p$ represents the output vector.

Functions f and h are vector systems assumed to be continuously differentiable on x.

The input u is locally bounded and measurable.

The sliding mode observer is defined with the following structure [48]:

$$\begin{cases} \dot{\hat{x}} = f(\hat{x}, u) - K\text{sign}(\hat{y} - y) \\ \hat{y} = h(\hat{x}) \end{cases} \tag{72}$$

with K is the gain matrix of $(n - p)$ dimension.

The obtained observer is a copy of the system model plus a correction term which establishes the convergence of $\hat{x}$ to x (Fig. 2).

The sliding surface in this case is given by:

$s(x) = y - \hat{y}$.

The correction term used is proportional to the discontinuous signum function applied to the output error that is defined by [48]:

$$sign(x) = \begin{cases} 1 \ if \ x > 0 \\ 0 \ if \ x = 0 \\ -1 \ if \ x < 0 \end{cases} \tag{73}$$

The sliding mode observer must respect two conditions in order to guarantee that the estimated state converge to the real state:

- The first condition concerns the reaching mode and guarantees the attractiveness of the sliding surface $S = 0$ with p dimension. The sliding surface is attractive if the Lyapunov function $V(x) = S^T \times S$ verifies the condition: $\dot{V}(x) < 0$ if $S \neq 0$
- The second one, concerns the sliding mode. During this step, the corrective gain matrix satisfies the following invariance condition:

$$\begin{cases} \dot{S} = 0 \\ S = 0 \end{cases} \tag{74}$$

The system dynamics are reduced and the n-order system becomes an equivalent $(n - p)$ order system. These criteria allow the synthesis of the sliding mode observer and determine its operation [49].

6.2 *Sliding Mode Observer of Linear Systems*

Considering the following linear system:

$$\begin{cases} \dot{x} = Ax(t) + Bu(t) \\ y = Cx(t) \end{cases} \tag{75}$$

where $x \in \mathbb{R}^n$ is the state vector, $u \in \mathbb{R}^m$ is the vector inputs, $y \in \mathbb{R}^p$ denotes the output vector.

Matrices A, B and C have appropriate dimensions.

The pair (A, C) is assumed to be observable.

The reconstruction of the state variables is based on the measured outputs. A change of coordinates can be performed so that the outputs appear directly as components of the state vector.

Recalling Eq. (51), a non-singular transformation matrix T allows to rewrite respectively the output, state and control matrices as follows:

$$\tilde{A} = TAT^{-1} = \begin{bmatrix} A_{11} & A_{12} \\ A_{21} & A_{22} \end{bmatrix} \tag{76}$$

$$\tilde{B} = TB = \begin{bmatrix} B_1 \\ B_2 \end{bmatrix} \tag{77}$$

The linear system presented in Eq. (75) can thus be in the following form:

$$\begin{cases} \dot{x}_1(t) = A_{11}x_1(t) + A_{12}y(t) + B_1u(t) \\ \dot{y} = A_{21}x_1(t) + A_{22}y(t) + B_2u(t) \end{cases} \tag{78}$$

$$Tx(t) = \begin{bmatrix} x_1(t) \\ y(t) \end{bmatrix} \tag{79}$$

with $x_1(t) \in \mathbb{R}^{n-p}$ The proposed sliding mode observer for this type of system is expressed as:

$$\begin{cases} \dot{\hat{x}}_1(t) = A_{11}\hat{x}_1(t) + A_{12}\hat{y}(t) + B_1u(t) + LKsign(\hat{y}_i(t) - y_i(t)) \\ \dot{\hat{y}} = A_{21}\hat{x}_1(t) + A_{22}\hat{y}(t) + B_2u(t) - Ksign(\hat{y}_i(t) - y_i(t)) \end{cases} \tag{80}$$

with $L \in \mathbb{R}^{(n-p)\times p}$ is the observer gain, $K > 0$ and $\hat{y}_i(t)$ and $y_i(t)$ are the vector components of $\hat{y}(t)$ and $y(t)$, respectively.

The state and output estimation errors are given by :

$$\begin{cases} e_1(t) = \hat{x}_1(t) - x_1(t) \\ e_y(t) = \hat{y}(t) + y(t) \end{cases} \tag{81}$$

From Eqs. (78), (80) and (81), the dynamics of the estimation errors will be written as:

$$\begin{cases} \dot{e}_1(t) = A_{11}e_1(t) + A_{12}e_y(t) + LKsign(\hat{y}_i(t) - y_i(t)) \\ \dot{e}_y(t) = A_{21}e_1(t) + A_{22}e_y(t) - Ksign(\hat{y}_i(t) - y_i(t)) \end{cases} \tag{82}$$

The pair (A_{11}, A_{21}) is observable because the pair (A, C) is observable. Therefore, the gain L can be chosen such that the eigenvalues of the matrix $A_{11} + LA_{21}$ are in the left half-plane plane of the complex plane.

6.3 Triangular Sliding Mode Observer

The triangular sliding mode observer has the following form:

$$\begin{cases} \begin{pmatrix} \dot{x}_1 \\ \dot{x}_2 \\ \cdot \\ \cdot \\ \cdot \\ \dot{x}_{n-1} \\ \dot{x}_n \end{pmatrix} = \begin{pmatrix} x_2 + g_1(x_1, u) \\ x_3 + g_2(x_1, x_2, u) \\ \cdot \\ \cdot \\ \cdot \\ x_n + g_{n-1}(x_1, x_2, \ldots, u) \\ f_n(x) + g_n(x, u) \end{pmatrix} \\ y = x_1 \end{cases} \tag{83}$$

where g_i and f_n for $i = 1, 2, \ldots, n$ are the analytic functions, $x = [x_1 x_2 \ldots x_n]^T \in \mathbb{R}^n$ is the system state, $u \in \mathbb{R}^m$ is the input vector and $y \in \mathbb{R}$ is the output.

The proposed observer structure is as follows:

$$\begin{cases} \begin{pmatrix} \dot{\hat{x}}_1 \\ \dot{\hat{x}}_2 \\ \cdot \\ \cdot \\ \cdot \\ \dot{\hat{x}}_{n-1} \\ \dot{\hat{x}}_n \end{pmatrix} = \begin{pmatrix} \hat{x}_2 + g_1(x_1, u) + \lambda_1 \text{sign}_1(x_1 - \hat{x}_1) \\ \hat{x}_3 + g_2(x_1, \bar{x}_2, u) + \lambda_2 \text{sign}_2(\bar{x}_2 - \hat{x}_2) \\ \cdot \\ \cdot \\ \cdot \\ \hat{x}_n + g_{n-1}(x_1, \bar{x}_2, \ldots, \bar{x}_{n-1}, u) + \lambda_{n-1} \text{sign}_{n-1}(\bar{x}_{n-1} - \hat{x}_{n-1}) \\ f_n(x_1, \bar{x}_2, \ldots, \bar{x}_n) + g_n(x_1, \bar{x}_2, \ldots, \bar{x}_n, u) \lambda_n \text{sign}_n(\bar{x}_n - \hat{x}_n) \end{pmatrix} \\ \hat{y} = \hat{x}_1 \end{cases} \tag{84}$$

where $\bar{x}_i = \hat{x}_i + \lambda_{i-1}\text{sign}_{moy,i-1}(\bar{x}_{i-1} - \hat{x}_{i-1})$ with $\text{sign}_{moy,i-1}$ denoting the function sign_{i-1} filtered by a low pass filter. $\text{sign}_i(.)$ is equal to zero if there exists $j \in \{1, \ldots, i-1\}$ such that $\bar{x}_j - \hat{x}_j \neq 0$ (by definition $\bar{x}_1 = x_1$), if not $\text{sign}_i(.)$ is taken equal to the classical function sign(.). According to these propositions, we impose that the corrector term is "active" only if the condition $\bar{x}_j - \hat{x}_j = 0$ for $j = 1, 2, \ldots, i-1$ is verified.

There exists a choice of λ_j such that the observer state $\hat{x}$ converges in a finite time to the state x of the system.

Let us consider the dynamics of the observer error $e = x - \hat{x}$ and proceed step by step. For $e_1 = x_1 - \hat{x}_1$, we obtain: $\dot{e}_1 = e_2 - \lambda_1\text{sign}(e_1)$ with $e_2 = x_2 - \hat{x}_2$.

If $\lambda_1 > |e_2|_{max}$ for $t > t_1$, then the sliding surface $e_1 = 0$ is reached after a finite time t_1 which means that $\dot{e}_1 = 0$.

There is a continuous function noted sign_{eq} defined by: $e_2 - \lambda_1\text{sign}_{eq}(e_1) = 0$, involving $\bar{x}_2 = x_2$ on the sliding surface, since $\text{sign}_{eq} = \text{sign}_{moy}$, then:

$$\dot{e}_1 = x_2 - (\hat{x}_2 + \lambda_1\text{sign}_{eq}(x_1 - \bar{x}_1)) = x_2 - \bar{x}_2 = 0 \tag{85}$$

Once x_2 is known, we will move on to the dynamics of e_2.

After, t_1, we obtain $\bar{x}_2 = x_2$ which implies that: $g_1(x_1, x_2) - g_2(x_1, \bar{x}_2) = 0$. Then, $\dot{e}_2 = e_3 - \lambda_2\text{sign}(e_2)$. Following the same reasoning, if $\lambda_2 > |e_3|_{max}$ for $t > t_2$, we will obtain after a finite time $t_2 > t_1$, the convergence to the surface $e_1 = e_2 = 0$. The dynamics of the remaining observer error on the sliding surface is given by $\dot{e}_2 = 0$. Then, $x_3 = \bar{x}_3$ because: $\dot{e}_2 = x_3 - (\hat{x}_3 + \lambda_2\text{sign}_{eq}(x_2 - \bar{x}_2) = x_3 - \bar{x}_3 = 0$.

By reiterating $(n-1)$ times this process, we obtain after t_{n-1} convergence of all the observer errors on the sliding surface $e_1 = e_2 = \cdots = 0$ and consequently $\bar{x}$ tends towards x, in a finite time t_{n-1} all the state is known and the observer error is zero.

7 Conclusion

Preliminaries on unknown input observers are presented in this chapter. The evolution of discontinuous observers, as well as the methods by which they are constructed, are also discussed. The sliding mode observers are the focus of the chapter's concluding sections.

References

1. Zhu, Z., Leung, H.: Identification of linear systems driven by chaotic signals using nonlinear prediction. IEEE Trans. Circuits Syst. I : Fund. Theory Appl. **49**(2), 170–180 (2002)
2. Williams, H.F.: A solution of the multivariable observer for linear time varying discrete systems. In: Rec. 2nd Asilomar Conference Circuit and systems, pp. 124–129 (1968)
3. Kalman, R.E., Betram, J.E.: Control system analysis and design via the "second method" of Lyapunov -I : continuous-time system. ASME J. Basic Eng. **82**, 371–393 (1960)
4. Aoki, M., Huddle, J.R.: Estimation of the state vector of a linear stochastic system with a constrained estimator. IEEE Trans. Autom. Control **12**(4), 432–433 (1967)
5. Newmann, M.M.: Specific optimal control of the linear regulator using a dynamical controller based on the minimal-order observer. Int. J. Control **12**, 33–48 (1970)
6. Raymond T.S.: Observer steady-state errors induced by errors in realization. IEEE Trans Autom Control 280–281 (1976)
7. Busawon, K.K., Kabor, P.: On the design of integral and proportional integral observers. In: American Control Conference, vol. 6, pp. 3725–3729 (2000)
8. Koenig, D., Mammar, S.: Design of proportional-integral observer for unknown input descriptor systems. IEEE Trans. Autom. Control **47**(12), 2057–2062 (2002)
9. Gao, Z., Ho, D.W.C.: Proportional multiple-integral observer design for descriptor systems with measurement output disturbances. In: IEEE Proceedings Control Theory and Applications, vol. 151, pp. 279–288 (2004)
10. Johnson, C.D.: Observers for linear systems with unknown and inaccessible inputs. Int. J. Control **21**, 825–831 (1975)
11. Meditch, J.S., Hostetter, G.H.: Observers for systems with unknown and inaccessible inputs. Int. J. Control **19**, 637–640 (1971)
12. Xiong, Y., Saif, M.: Unknown disturbance inputs estimation based on a state functional observer design. Automatica **39**, 1389–1398 (2003)
13. Dassanake, S.K., Balas, G.L., Bokor, J.: Using unknown input observers to detect and isolate sensor faults in a turbofan engine. In: Digital Avionics Systems Conferences, vol. 7, pp. 6E51–6E57 (2000)
14. Valcher, M.E.: State observers for discrete-time linear systems with unknown inputs. IEEE Trans. Autom. Control **44**(2), 397–401 (1999)
15. Wang, S.H., Davison, E.J., Dorato, P.: Observing the states of systems with unmeasurable disturbances. IEEE Trans. Autom. Control **20**, 716–717 (1975)
16. Moreno, J.: Quasi-Unknown input observers for linear systems. In: IEEE Conference on Decision and Control, pp. 732–737 (2001)
17. Kobayashi, N., Nakamizo, T.: An observer design for linear systems with unknown inputs. Int. J. Control **35**, 605–619 (1982)
18. Lyubchik, L.M., Kostenko, Y.T.: The output control of multivariable systems with unmeasurable arbitrary disturbances-the inverse model approach. In: ECC'93, pp. 1160–1165. Groningen, The Netherlands (1993)
19. Liu, C.S.., Peng, H.: Inverse-dynamics based state and disturbance observers for linear time-invariant systems. J. Dyn. Syst. Meas. Control **124**, 375–381 (2002)

20. Kudva, P., Viswanadham, N., Ramakrishna, A.: Observers for linear systems with unknown inputs. IEEE Trans. Autom. Control **25**(1), 113–115 (1980)
21. Guan, Y., Saif, M.: A novel approach to the design of unknown input observers. IEEE Trans. Autom. Control **36**(5), 632–635 (1991)
22. Yang, F., Wilde, R.WW.: Observers for linear systems with unknown inputs. IEEE Trans. Autom. Control **33**, 677 (1988)
23. Zasadzinski, M., Magarotto, E., Darouach, M.: Unknown input reduced order observer for singular bilinear systems with bilinear measurements. In: IEEE Conference on Decision and Control, vol. 1, pp. 796–801 (2000)
24. Koenig, D., Mammar, S.: Design of a class of reduced order unknown inputs nonlinear observer for fault diagnosis. In: American Control Conference, vol. 3, pp. 2143–2147 (2001)
25. Watanabe, K., Himmelblau, D.M.: Instrument fault detection in systems with uncertainties. Int. J. Syst. Sci. **13**, 137 (1982)
26. Saif, M., Guan, Y.: A new approach to robust fault detection and identification. IEEE Trans. Aerosp Electron Syst **29**(3), 685–695 (1993)
27. Guerra, R.M., Diop, S.: Diagnosis of nonlinear systems using an unknowninput observer : an algebraic and differential approach. IEEE Control Theory Appl. **15**(1), 130–135 (2004)
28. Hou, M., Mtiller, P.C.: Design of observers for linear systems with unknown inputs. IEEE Trans. Autom. Control **37**, 871 (1992)
29. Darouach, M., Zasadzinski, M., Xu, S.J.: Full-order observers for linear systems with unknown inputs. IEEE Trans. Autom. Control **39**(3), 606–609 (1994)
30. Chen, J., Zhang, H.: Robust detection of faulty actuators via unknown input observers. Int. J. Syst. Sci. **22**, 1829–1839 (1991)
31. Rotella, F., Borne, P.: The téorie et pratique du calcul matriciel, Editions Technip (1995)
32. Chadli, M., Maquin, D., Ragot, J.: Output stabilization in multiple model approach. Int. Conf. Control Appl. **2**, 1315–1320 (2002)
33. Boyd, S., El Ghaoui, L., Feron, E., Balakrishnan, V.: Linear Matrix Inequalities in System and Control Theory. SIAM, Philadelphia (1994)
34. Utkin, V.I.: Sliding Modes in Control Optimization. Springer, Berlin (1992)
35. DeCarlo, R.A., Zak, S.H., Matthews, G.P.: Variable structure control of nonlinear multivariable systems : a tutorial. Proc. IEEE **76**(3), 212–232 (1988)
36. Utkin, V.I.: Principles of identification using sliding regimes. Soviet Trans. Autom. Control **26**, 271–272 (1981)
37. Dorling, C.M., Zinober, A.S.I.: Two approaches to hyperplane design in multivariable variable structure control systems. Int. J. Control **44**, 65–82 (1986)
38. Walcott, B.L., Corless, M.J., Zak, S.H.: Comparative study of nonlinear state observation. Int. J. Control **45**, 2109–2132 (1987)
39. Walcott, B.L., Zak, S.H.: Combined observer-controller synthesis for uncertain dynamical systems with applications. IEEE Trans. Syst. Man Cybern. 18, 88–104 (1988)
40. Zak, S.H., Walcott, B.L.: State observation of nonlinear control systems via the method of Lyapunov. In: Zinober, A.S.I. (ed.) Deterministic Control of Uncertain Systems. Peter Peregrinus, London, UK (1990)
41. Ha, Q.P., Nguyen, H.T., Tuan, H.D.: Dynamic output feedback sliding-mode control using pole placement and linear functional observers. IEEE Trans. Indus.Electron. **50**(5), 1030–1037 (2003)
42. Xiong, Y., Saif, M.: Sliding mode observer for nonlinear uncertain systems. IEEE Trans. Autom. Control **46**(12), 2012–2017 (2001)
43. Islam, M.S., Husain, I., Veillette, R.J., Batur, C.: Sliding mode observer for nonlinear uncertain systems. IEEE Trans. Control Syst. Technol. **11**(3), 383–389 (2003)
44. Walcott, B.L., Corless, M.J., Zak, S.H.: Obervation of dynamical systems in the presence of bounded nonlinearitie/uncertainties. In: IEEE Conference on Decision and Control, pp. 961–966 (1986)
45. Edwards, C., Spurgeon, S.K.: On the development of discontinuous observers. Int. J. Control **25**, 1211–1229 (1994)

46. Edwards, C., Spurgeon, S.K.: Sliding mode observers for fault detection and isolation. Automatica **36**(4), 541–553 (2000)
47. Slotine, J.J.E., Hedrick, J.K., Misawa, E.A.: Nonlinear state estimation using sliding observers. In: 25th IEEE Conference, pp. 332–339 (1986)
48. Sira-Ramirez, H.: On the sliding mode control of nonlinear systems. Syst. Control Lett. 303–312 (1992)
49. Solvar, S.: Observateur à mode glissant d'ordre 2 appliqué à la MAS sans capteur mécanique, réunion GDR CE2, E.N.S.A.M (2010)

Observers with Unknown Inputs of Linear Systems

Dorsaf Etlili, Omar Naifar, and Ayachi Errachdi

Abstract This chapter introduces the concepts of observability and observer for linear systems, as well as the concept of sliding mode. It is demonstrated how to build an observer for a linear dynamical system with unknown inputs. The determination of the observer's gain in order to ensure convergence leads to the solution of an LMI issue (bilinear matrix inequalities). The technique based on a change of variables is used to resolve these LMI restrictions, allowing the matrices characterizing the observer to be determined.

1 Introduction

The complete or partial knowledge of the state of the considered system is an important requirement in the fields of control, diagnosis and monitoring of systems. This requirement is difficult to satisfy in most cases. This is due, on the one hand, to the fact that the state variables do not always have a physical meaning and their direct measurement is impossible to achieve. On the other hand, the sensors needed to measure the state variables are unavailable or of insufficient accuracy. Moreover, from an economic point of view, it is desirable to install a minimum of sensors in order to reduce the costs of instrumentation and maintenance.

The measurements made at the output of the system do not give complete information on the internal states of this system. It is therefore essential to reconstruct the unmeasured state variables. The idea used for several years, is the replacement of

D. Etlili
University of Tunis El Manar, National School of Engineering of Tunis, Tunis, Tunisia

O. Naifar (✉)
Control and Energy Management Laboratory, University of Sfax, National School of Engineering of Sfax, Sfax, Tunisia

A. Errachdi
University of Kairouan, Higher Institute of Apllied Sciences and Technology of Kairouan, Kairouan, Tunisia

O. Naifar and A. Ben Makhlouf (eds.), *Advances in Observer Design and Observation for Nonlinear Systems*, Studies in Systems, Decision and Control 410,
https://doi.org/10.1007/978-3-030-92731-8_4

hardware sensors by software sensors or by state observers, which allow to reconstruct the internal information (states, unknown inputs) of the system from the system model, the known inputs and the measured outputs.

A physical system is often subject to disturbances, such as measurement noise, measurement uncertainties, system faults and external disturbances. These noises have an adverse effect on the normal behavior of the process, and their estimation can be used to design a control system capable of minimizing these effects. These disturbances are called unknown inputs when they appear as additional inputs to the process, and their presence can make the estimation of system states difficult.

Several works have been devoted to the synthesis of observers for linear systems with unknown inputs [1–7] The first results on linear state estimation date back to the 1970's. They can be grouped into two categories. The first category assumes a priori knowledge of information about these non-measurable inputs; in particular, Johnson [1] proposes a polynomial approach and Meditch [8] suggests to approximate the unknown inputs by the response of a known dynamical system. The second category proceeds either by estimating the unknown input [7] or by its complete elimination from the system equations [9].

Reduced order observers have been considered by several authors during the last years [10, 11]. However, Yang and Wilde [10] demonstrated that the full order unknown input observer can have a faster convergence speed than the reduced order observer.

The use of observers with unknown inputs for fault diagnosis and process monitoring systems has also attracted much attention [9, 11, 12]. In this chapter, we present some basic notions of observability and observers as well as some methods for reconstructing the states and unknown input of linear systems in the presence of unknown input.

2 Observability

In the literature, it is shown that an observer exists if and only if the state realization of the system in question is observable. Indeed, the observability of a system expresses the possibility of reconstructing the state from the sole knowledge of the input and output signals.

3 Observability of Linear Systems

The observability criteria of a linear system are described in many references [13, 14]. Let us consider the continuous linear time-invariant dynamical system:

$$\begin{cases} \dot{x}(t) = Ax(t) + Bu(t) \\ y(t) = Cx(t) \end{cases} \tag{1}$$

where t ≥ 0; x(t) ∈ $\mathbb{R}^n$; u(t) ∈ $\mathbb{R}^m$; y(t) ∈ $\mathbb{R}^p$, are the state vector, the input and the output of the system (1), respectively. A, B, C are the matrices of the system of appropriate dimensions, and the initial conditions are defined by$x(t_0) = x_0$: Let us recall some definitions and results on observability.

Definition 1 (*Borne* [14]) The system (1) is said to be observable if there exists a time $t_f \geq t_0$ such that the knowledge of the input $u(t)$ and the output $y(t)$ on the interval t ∈$\left[t_0 t_f\right]$ is sufficient to determine the initial condition x_0 in a unique way.

For linear systems, the information produced at the output is the superposition of that generated by the input and that generated by the initial condition. If we assume the free regime (u = 0) then we can adopt the following definition.

Definition 2 (*Borne* [14]) The system (1) is observable if and only if, in the free regime ($u(t) = 0$; ∀t≥ t_0), the observation of a uniformly zero output $y(t)\, t \in \left[t_0\, t_f\right]$ is possible only for an initial state $x(t_0)$ zero.

Remarks 1 When all state variables are observable, then the system is said to be completely observable, otherwise it is said to be partially observable.

The observability condition is a necessary and sufficient condition to be able to estimate the state of the system from the information collected on the inputs and outputs. Note that the knowledge of x_0 and the state model of the system is sufficient to reconstruct the state x(t) at any time t ≥ t_0. The observability property of a linear time invariant system is a structural property and depends only on the matrices A and C of the model. The most used criterion to check this property is the Kalman rank criterion formulated by the observability matrix below.

The system described by (1) is completely observable if and only if rank (O) = n such that (O) is the observability matrix defined by:

$$O = \begin{bmatrix} C \\ CA \\ \cdot \\ \cdot \\ \cdot \\ CA^{n-1} \end{bmatrix} \tag{2}$$

4 Synthesis of Observers for Linear Systems Without Unknown Inputs

A solution to the problem of state estimation of linear systems has been proposed by Luenberger [15] in the deterministic framework and by Kalman [16] in the stochastic framework. Sliding mode observers are also used for linear systems even if they are themselves of nonlinear structure.

4.1 Luenberger Observer

The theory of observation is essentially based on pole placement techniques. Let $\widehat{x}(t)$ be the estimate of x(t); and $\widehat{y}(t)$ the estimate of y(t).

The observer proposed by Luenberger for the system (1) is described by the following equations:

$$\begin{cases} \dot{\hat{x}}(t) = A\hat{x}(t) + Bu(t) + K(y(t) - \hat{y}(t)); \hat{x}(t_0) = \hat{x}_0 \\ \hat{y}(t) = C\hat{x}(t) \end{cases} \tag{3}$$

where $K \in \mathbb{R}^{n*p}$ is the gain of the observer (3). The block diagram of the observer is illustrated by Fig. 1. The estimation error is given by

$$e(t) = x(t) - \hat{x}(t)$$

The dynamics of this error is governed by the following equation:

$$\dot{e}(t) = (A - KC)e(t); e(t_0) = e_0 = x_0 - \hat{x}_0$$

If the gain is chosen such that the matrix (A−KC) is Hurwitz, i.e., has strictly negative eigenvalues, then the estimation error converges asymptotically to zero. As the observer replaces the sensor, we must therefore ensure a convergence of the estimation error to zero very fast, at least ten times faster than the dynamics of the system. If the couple (A, C) is observable, then it is possible to determine the gain K to have a convergence dynamics chosen beforehand. The problem of constructing the observer is therefore equivalent to solving a pole placement problem. We choose a desired dynamics (choice of the desired eigenvalues of (A−KC), then using the pole placement principle, we determine the gain K.

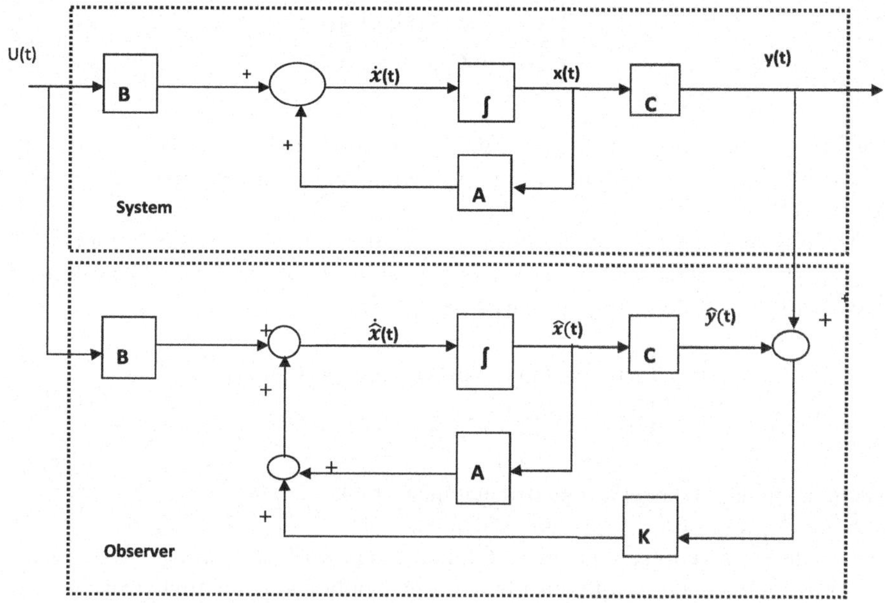

Fig. 1 Block diagram of the Luenberger observer

4.2 Sliding Mode Observer

Techniques based on the theory of variable structures, ensuring the robustness of the controller or the observer, are currently the subject of several research works. One of the best known classes of robust observers is that of sliding mode observers [17–20].

In [18], the principle of sliding mode observers consists in constraining, by means of discontinuous functions, the dynamics of a system of order n to converge to a sliding surface S of dimension (n−p) (p being the dimension of the measurement vector y). The attractiveness of this surface is ensured by conditions called sliding conditions. If these conditions are satisfied, the system converges towards the sliding surface and evolves there according to a dynamics of order (n-p). In the case of sliding mode observers, the dynamics concerned are those of the observation errors $e(t) = x(t) - \hat{x}(t)$. From their initial values e_0, these errors converge to the equilibrium values in two steps:

In the first stage, the trajectory of the observation errors evolves towards the sliding surface on which the errors between the observer output and the real system output (the measurements) $e_y = y - \hat{y}$ are zero. This stage is called the attainment mode.

In the second phase, the trajectory of the observation errors slides on the sliding surface with imposed dynamics so as to cancel all observation errors. This last mode is called sliding mode.

Consider a nonlinear state system of order n:

$$\begin{cases} \dot{x}(t) = f(x(t), u(t)) \\ y(t) = Cx(t) \end{cases} \tag{4}$$

where: x(t) $\in \mathbb{R}^n$; u(t) $\in \mathbb{R}^m$; y(t) $\in \mathbb{R}^p$, f represent the state vector, the input or control vector, the output vector, the sufficiently differentiable vector field, respectively.

The different steps for the synthesis of the sliding mode observer are identified in [17]. The first order sliding mode observer allowing to reconstruct the estimated state vector $\hat{x}(t)$ is defined by the structure (5)

$$\begin{cases} \dot{\hat{x}}(t) = f(\hat{x}(t), u(t)) + \lambda sign(y - \hat{y}) \\ \hat{y}(t) = C\hat{x}(t) \end{cases} \tag{5}$$

where the input u is locally bounded and measurable.

Sign: Represents the usual sign function.

λ: is the observation gain matrix of dimension (n–p). The correction term used is proportional to the discontinuous function sign applied to the output error.

For the estimated state to converge to the true state, the sliding mode observer must satisfy two conditions:

The first condition concerns the mode of reaching and guarantees the attractiveness of the sliding surface S = 0 of dimension p.

The sliding surface is attractive if the Lyapunov function $V(t) = S^T S$ satisfies the condition: $\dot{V}(t) \prec 0$.

The second one concerns the sliding mode, during this step, the corrective gain matrix acts so as to satisfy the following invariance condition:

$$\begin{cases} S = 0 \\ \dot{S} = 0 \end{cases}$$

During this mode, the dynamics of the system are reduced and the system of order n becomes an equivalent system of order (n–p). These criteria allow the synthesis of the sliding mode observer and determine its operation.

Phenomenon of reluctance

In practice, the discontinuous term on the right-hand side of the equation can excite unmodelled high-frequency dynamics that lead to the appearance of what is known as "reticence" or "chattering", which is characterized by strong oscillations around the surface.

5 Synthesis of Observers for Linear Systems with Unknown Inputs

5.1 *Utkin Sliding Mode Observer with Unknown Input*

Let us consider the continuous linear system time invariant with delay on the measurement

$$\begin{cases} \dot{x}(t) = Ax(t) + Bu(t) + Rw(t) \\ y(t) = Cx(t) \end{cases} \tag{6}$$

x(t) $\in \mathbb{R}^n$; u(t) $\in \mathbb{R}^m$; y(t) $\in \mathbb{R}^p$, $w(t) \in \mathbb{R}^q$ Are the state vectors, the vector of known inputs, the vector of measurable outputs, the vector of unknown inputs of the system (6), respectively. $A \in \mathbb{R}^{n*n}$; B $\in \mathbb{R}^{n*m}$; C $\in \mathbb{R}^{p*n}$, $R \in \mathbb{R}^{n*q}$ are the state matrix, the matrix of known inputs, the influence matrix of unknown inputs and the output matrix of the system (6), respectively. It is assumed that R is of full column rank and the pair (A; C) is observable. The reconstruction of the state variables is based on the measured outputs; a coordinate change can be performed to obtain the regular form [21].

By respecting these conditions a non-singular transformation matrix allows to rewrite respectively the output, state and control matrices in the new coordinates.

$$\tilde{A} = T_1 A T_1^{-1} = \begin{pmatrix} A_{11} & A_{12} \\ A_{21} & A_{22} \end{pmatrix}, \tilde{B} = T_1 B = \begin{bmatrix} B_1 \\ B_2 \end{bmatrix}, \tilde{C} = C T_1^{-1}, \tilde{R} = T_1 R = \begin{bmatrix} R_1 \\ R_2 \end{bmatrix}$$

où

$$T_1 = \begin{bmatrix} Q \\ C \end{bmatrix}$$

$$\begin{bmatrix} \tilde{x}_1(t) \\ y(t) \end{bmatrix} = T_1 x(t), \tilde{x}_1(t) \in \mathbb{R}^{(n-p)}, R = \begin{bmatrix} R_1 \\ R_2 \end{bmatrix}$$

where I_p is the identity matrix of dimension p.

$Q = \begin{bmatrix} 0 & I_{n-p} \end{bmatrix}$ the system (6) in the new coordinates is written as follows

$$\begin{cases} \dot{\tilde{x}}_1(t) = A_{11}\tilde{x}_1(t) + A_{12}y(t) + B_1 u(t) + R_1 w(t) \\ \dot{y}(t) = A_{21}\tilde{x}_1(t) + A_{22}y(t) + B_2 u(t) + R_2 w(t) \end{cases} \tag{7}$$

We note that $CR = R_2$: so $CR \neq 0$ and there exists the pseudo-inverse matrix R_2^+ of the matrix R_2 such that $R_2 R_2^+ = I_{m_1}, m_1 = rang(CR) = rang(R)$.

The following transformation is applied to the model given in (7)

$$\begin{bmatrix} \overline{x}_1(t) \\ y(t) \end{bmatrix} = \begin{bmatrix} I_{n-p} & -R_1 R_2^+ \\ 0 & I_p \end{bmatrix} \begin{bmatrix} \tilde{x}_1(t) \\ y(t) \end{bmatrix}$$

with

$$T_2 = \begin{bmatrix} I_{n-p} & -R_1 R_2^+ \\ 0 & I_p \end{bmatrix}, \overline{A} = T_2 \tilde{A} T_2^{-1} = \begin{bmatrix} \overline{A}_{11} & \overline{A}_{12} \\ \overline{A}_{21} & \overline{A}_{22} \end{bmatrix}, \overline{B} = T_2 \tilde{B} = \begin{bmatrix} \overline{B}_1 \\ \overline{B}_2 \end{bmatrix}$$

$$\overline{C} = \tilde{C} T_2^{-2}, \overline{R} = T_2 \tilde{R} = \begin{bmatrix} \overline{R}_1 \\ \overline{R}_2 \end{bmatrix}$$

Where

$\overline{R}_1 = 0$

$-R_1 R_2^+$ is the pseudo-inverse of $-R_1 R_2$ and $x_1(1) \in R^{n-p}$.
The system (7) in the new coordinates is given by (8)

$$\begin{cases} \dot{\overline{x}}_1(t) = \overline{A}_{11}\overline{x}_1(t) + \overline{A}_{12}y(t) + \overline{B}_1 u(t) \\ \dot{y}(t) = \overline{A}_{21}\overline{x}_1(t) + \overline{A}_{22}y(t) + \overline{B}_2 u(t) + \overline{R}_2 w(t) \end{cases} \tag{8}$$

The pair ($\overline{A}_{11}$;$\overline{A}_{21}$) is observable because the pair (A; C) is observable [22], the gain L is chosen such that the eigenvalues of the matrix $\overline{A}_{11} - L\overline{A}_{21}$ are in the left half plane of the complex plane.

5.2 *Structure of the Utkin Sliding Mode Observer with Unknown Input*

The sliding mode observer structure considered for this system is:

$$\begin{cases} \dot{\hat{x}}_1(t) = \overline{A}_{11}\hat{x}_1(t) + \overline{A}_{12}\hat{y}(t) + \overline{B}_1 u(t) + \overline{L}v(t) \\ \dot{\hat{y}}(t) = \overline{A}_{21}\hat{x}_1(t) + \overline{A}_{22}\hat{y}(t) + \overline{B}_2 u(t) + Iv(t) \end{cases} \tag{9}$$

where $\hat{y}(t)$ and $\hat{x}_1(t)$ are the estimates of $y(t)$ and $x_1(t)$ respectively, L is the observer gain and $v(t)$ is the discontinuous function given by:

$$v(t) = Msign(\hat{y}(t) - y(t))$$

With M > 0. The state and output estimation errors

$$\begin{cases} e_1(t) = \hat{x}_1(t) - x_1(t) \\ e_y(t) = \hat{y}(t) - y(t) \end{cases} \tag{10}$$

Subtracting (8) from (9), the dynamics of the estimation errors are written as follows:

$$\begin{cases} \dot{e}_1(t) = \overline{A}_{11}e_1(t) + \overline{A}_{12}e_y(t) + \overline{L}v(t) \\ \dot{e}_y(t) = \overline{A}_{21}e_1(t) + \overline{A}_{22}e_y(t) - v(t) + R_2w(t) \end{cases} \tag{11}$$

We perform the following change of variable

$$\begin{bmatrix} \tilde{e}_1(t) \\ e_y \end{bmatrix} = \begin{bmatrix} I_{n-p} & -\overline{L} \\ 0 & I_p \end{bmatrix} \begin{bmatrix} e_1(t) \\ e_y \end{bmatrix}$$

The dynamics of the estimation errors will be written as follows:

$$\begin{cases} \dot{\tilde{e}}_1(t) = \tilde{A}_{11}\tilde{e}_1(t) + \tilde{A}_{12}e_y(t) + \overline{LR_2}w(t) \\ \dot{e}_y(t) = \tilde{A}_{21}\tilde{e}_1(t) + \tilde{A}_{22}e_y(t) - v(t) + \overline{R}_2w(t) \end{cases} \tag{12}$$

$\tilde{e}_1(t) = \tilde{e}_1(t) + Le_y(t)$ and $\tilde{A}_{11} = \overline{A}_{11} - L\overline{A}_{21}$.
$\tilde{A}_{12} = \overline{A}_{12} - L\overline{A}_{22} + \tilde{A}_{11}\overline{L}$ and $\tilde{A}_{22} = \overline{A}_{22} - \overline{A}_{21}\overline{L}$.

Utkin [21] has shown using the theory of singular perturbations, for a large enough gain M the sliding regime can be established on the error (12). So after a finite time the error e(t) and its derivative will be zero and we have from Eq. (12).

$$\dot{\tilde{e}}_1(t) = \tilde{A}_{11}e_1(t)$$

The gain M is chosen such that $\tilde{A}_{11}$ is stable and thus the system of Eqs. (12) converges asymptotically to zero, and $\tilde{e}_1(t) \to 0$ when $t \to \infty$.

The equivalent control method is used to obtain the estimated unknown input. It is assumed that the error of the system (12) is in the slip along $e_y = 0$: thus $\dot{\tilde{e}}_1 = 0$ and $\dot{\tilde{e}}_y = 0$. The solution of the system of Eq. (12) for $w(t)$ gives us the following estimate of $w(t)$:

$$\hat{w} \approx \left((I + \tilde{A}_{21}\tilde{A}_{11}^{-1}\tilde{L})\tilde{R}_2\right)^+ v_{eq} \tag{13}$$

where v_{eq} is the equivalent command.

6 Luenberger Observer with Unknown Input

In this section, we present the synthesis of large gain type observers for the class of uniformly observable nonlinear systems to which we have added unknown inputs. The proposed observers jointly estimate the entire state of the system as well as all unknown inputs under sufficient conditions that will be given. Their synthesis neither assumes nor adopts any mathematical model for the unknown inputs. We simply assume that the first derivative with respect to time of each of the unknown inputs is bounded.

Before presenting the class of nonlinear systems that will be the object of our study, we propose to recall the necessary and sufficient conditions for the synthesis of an observer with unknown inputs for linear systems. This will allow us to better understand the sufficient conditions that we will adopt for the synthesis of the proposed observers.

Note the necessary and sufficient conditions that will be recalled for linear systems concern the synthesis of an observer allowing the estimation of states (via a full or reduced order observer) without any knowledge about the unknown inputs. These conditions can be relaxed if certain assumptions about these inputs, such as the boundedness of their derivatives with respect to time, are adopted. We will come back to this point later in this part.

- **Reminders on observers with unknown input synthesis for linear systems**:

We consider the following linear time invariant system:

$$\sum \begin{cases} \dot{x} = Ax + Bu + Gv \\ y = Cx \end{cases} \tag{14}$$

where state x(t) $\in \mathbb{R}^n$, known input u(t) $\in \mathbb{R}^\mu$, unknown input v(t) $\in \mathbb{R}^m$, output y $\in \mathbb{R}^p$, A, B, G, and C are known constant matrices of appropriate dimensions, and matrix G is assumed to be full rank in columns, i.e.

$$Rang(G) = m \tag{15}$$

Without detracting from generality, we assume that the matrix C has the following structure:

$$C = \begin{bmatrix} I_p \; 0 \; ... \; 0 \end{bmatrix} \tag{16}$$

In the same way, we will pose

$$G = \begin{pmatrix} G_1 \\ G_2 \end{pmatrix} \tag{17}$$

where $G_1 \in \mathbb{R}^{p*m}$ and $G_2 \in \mathbb{R}^{(n-p)\times m}$. Note that with this notation, we have

$$G_1 = CG \tag{18}$$

An observer with unknown inputs exists for this system if and only if the following two rank conditions are satisfied: [3–5].

$$Rang(CG) = m \tag{19}$$

$$Rang\begin{pmatrix} sI_n - A & G \\ C & 0 \end{pmatrix} = n + m, , \forall s \in C, \Re e(s) \geq 0 \tag{20}$$

We propose in the following to give some developments to show how these conditions are obtained. These developments will be mainly used to bring some complements on the synthesis of the observer when the number of unknown inputs is equal to the number of outputs.

The results we will present are described in [4] [3]. We will repeat them with more details here in the case where the matrix C has the particular structure (but not restrictive) (16).

The objective is to synthesize an observer that is written in the following form:

$$\Sigma\begin{cases} \dot{z} = Nz + Ly + Du \\ \hat{x} = z - Ey \end{cases} \tag{21}$$

where the observer state $z \in R^n$, $\hat{x} \in \mathbb{R}^n$ is the estimated state of the system x, $N \in \mathbb{R}^{n\times n}$, $L \in \mathbb{R}^{n\times p}$, $D \in \mathbb{R}^{n\times \mu}$, and $E \in \mathbb{R}^{n\times p}$ are matrices that must be chosen so that the observation error $e = \hat{x} - x$ converges asymptotically to 0.

To do this, let's pose

$$P = I_n + EC \tag{22}$$

The observation error is then written

$$\begin{aligned} e &= \hat{x} - x \\ &= z - Ey - x \\ &= z - (I_n + EC)x \\ &= z - Px \end{aligned}$$

It follows that

$$\begin{aligned}\dot{e} &= \dot{z} - P\dot{x} \\ &= Nz + Ly + Du - PAx - PBu - PGv \\ &= N(e + Px) + LCx - PAx - (PB - D)u - PGv \\ &= Ne + (NP - PA + LC)x - (PB - D)u - PGv\end{aligned} \tag{23}$$

If the matrices N, L, D and E are chosen so that the following conditions are satisfied

$$\text{N is Hurwitz matrix} \tag{24}$$

$$PG = (I_n + EC)G = 0 \tag{25}$$

$$D = PB = (I_n + EC)B \tag{26}$$

$$LC - PA = -NP \tag{27}$$

Then Eq. (23) becomes

$$\dot{e} = Ne \tag{28}$$

And the observation error converges asymptotically to 0.

It is now necessary to study under which conditions the choice of matrices N, L and E verifying (24), (25) and (27) is possible. Note that the matrix D is determined from E by the relation (26).

Equation (27) can be rewritten as follows

$$\begin{aligned}0 &= NP + LC - PA \\ &= N(I_n + EC) + LC - PA \Rightarrow \\ N &= PA - LC - NEC \\ &= PA - (L + NE)C \\ &= PA - KC\end{aligned} \tag{29}$$

$$K = L + NE \tag{30}$$

With

If we replace N by its expression (29) in Eq. (30), we obtain

$$K = L + (PA - KC)E$$

Or in an equivalent way

$$\begin{aligned} L &= K - (PA - KC)E \\ &= K(I_p + CE) - PAE \end{aligned} \tag{31}$$

The dynamics of the observer (21) then becomes:

$$\dot{z} = (PA - KC)z + Ly + Du \tag{32}$$

where the matrices P (or equivalently the matrix E), K, L and D are given by Eqs. (25), (30), (31) and (26). The problem of synthesizing the observer consists in finding a matrix E satisfying (25) and a vector K so that the matrix PA–KC is a Hurwitz matrix. This is a similar problem to that of the synthesis of classical observers. The eigenvalues of the matrix PA–KC can be chosen arbitrarily if and only if the pair (PA, C) is observable. Otherwise, a vector K such that the observation error (28) converges asymptotically to 0, exists if and only if the pair (PA, C) is detectable.

We will now discuss the conditions under which the matrix E (or equivalently the matrix P) exists.

Taking into account the particular structures considered for the C and G matrices (Eqs. (16) and (17)), Eq. (25) becomes

$$EG_1 = -G \tag{33}$$

The solution of Eq. (33) depends on the rank of the matrix G1 = CG. Note that since C is of rank plain with Rank C = p and Rank G = m, we have Rank G1 = min(p, m). There are two cases to consider:

1. $Rank(G_1) = p \prec m$
2. $Rank(G_1) = m \leq p$

Case 1: $Rank(G_1) = p \prec m$.

In this case, there is no solution for the matrix E. Indeed, the equality (33) cannot take place since on the one hand we have:

$$Rank(EG_1) \leq Rank(G_1) \prec m$$

And on the other hand, we have

$$Rank(-G) = Rank(G) = m$$

Since two equal matrices have trivially the same rank, the equation in E (33) does not admit any solution.

Case 2: $Rank(G_1) = m \leq p$.

The general solution of (33) is

$$E = -GG_1^{+} + Y(I_p - G_1G_1^{+}) \tag{34}$$

where G_1^+ is the left inverse of G_1 and $Y \in \mathbb{R}^{n \times p}$ is an arbitrary matrix. The matrix P can then be expressed as follows:

$$\begin{aligned} P &= I_n + EC \\ &= I_n - GG_1^+C + Y(I_P - GG_1^+)C \\ &= I_n + YC - GG_1^+C - YCGG_1^+C \\ &= I_n + YC - (I_n + YC)GG_1^+C \\ &= (I_n + YC)(I_n - GG_1^+C) \end{aligned} \quad (35)$$

Note that the maximum rank of the matrix,$n - m$, is obtained when the matrix $(I_n + YC)$ is non-singular [3].

We can now summarize the results obtained by thc following theorem [3]:

Theorem 1 *An observer of type* (32) *exists for the system* (14) *if and only if*:

(1) $Rank(CG) = Rank(G_1) = m$

(2) $Rank\begin{bmatrix} sI_n - PA \\ C \end{bmatrix} = n, \forall s \in \Re e(s) \geq 0$

We now give a second theorem which shows that the results of Theorem (14) *correspond to the conditions generally adopted for the synthesis of the observer* (32) [3]:

Theorem 2 *It is assumed that $Rank(CG) = Rank(G_1) = m$ and that $Rank(P) = n - m$. Then the following four conditions are equivalent*:

(1) The pair (PA,C) is detectable (observable);

(2) $Rank\begin{bmatrix} sP - PA \\ C \end{bmatrix} = n, \forall s \in \mathbb{C}; \Re e(s) \geq 0, (\forall s \in \mathbb{C});$

(3) $Rank\begin{bmatrix} sI_n - PA \\ C \end{bmatrix} = n, \forall s \in \mathbb{C}; \Re e(s) \geq 0;$

4) $Rank\begin{bmatrix} sI_n - A & G \\ C & 0 \end{bmatrix} = n + m, \forall s \in \mathbb{C}, \Re e(s) \geq 0, (\forall s \in \mathbb{C});$

We will now take a closer look at the case m = p, i.e. when the number of outputs is equal to the number of unknown inputs. This case has been discussed in [4] *and addressed in* [3]. *In this case we will give very simple conditions for the synthesis of the observer and we will give more details on the choice of the poles of the observer, when it exists.*

Special case: Rank(G1) = m and p = m.

We will look directly for the matrices N, L and E satisfying the conditions (24), (25) and (27).

Note that by multiplying each of the members of equality (27) on the right by G, and taking into account equality (25), we obtain

$$LCG = PAG$$

or equivalently, taking into account the structures of C and G,

$$LG_1 = PAG \tag{36}$$

Which then becomes the new equation fixing the choice of L.

As $m = p$ and the matrix G_1 is square and is invertible. From (33) and (36), we obtain:

$$E = -GG_1^{-1}$$
$$L = PAGG_1^{-1}$$

It remains now to study the choice of the matrix N. By noticing that:

$$\begin{aligned} LC - PA &= PAGG_1^{-1}C - PA \\ &= PA(GG_1^{-1}C - I_n) \\ &= -PA(EC + I_n) \end{aligned} \tag{37}$$

$$= -PAP \tag{38}$$

Equation (27) becomes

$$PAP = NP \tag{39}$$

Given the particular structures of C and G, the matrix

$$P = I_n - GG_1^{-1}C$$

Has the following topology

$$\begin{aligned} P &= \begin{bmatrix} I_m & 0 \\ 0 & I_{n-m} \end{bmatrix} - \begin{bmatrix} I_m & 0 \\ G_2G_1^{-1} & 0 \end{bmatrix} \\ &= \begin{bmatrix} 0 & 0 \\ -G_2G_1^{-1} & I_{n-m} \end{bmatrix} \end{aligned}$$

Now considering the following partitions of A and N:

$$A = \begin{bmatrix} A_{11} & A_{12} \\ A_{21} & A_{22} \end{bmatrix}; N = \begin{bmatrix} N_{11} & N_{12} \\ N_{21} & N_{22} \end{bmatrix}$$

where $A_{11}, N_{11} \in \mathbb{R}^{m\times m}$, $A_{12}, N_{12} \in \mathbb{R}^{m\times(n-m)}$, $A_{21}, N_{21} \in \mathbb{R}^{(n-m)\times m}$ and $A_{22}, N_{22} \in \mathbb{R}^{(n-m)\times(n-m)}$, we obtain

$$PAP = \begin{bmatrix} 0 & 0 \\ -(A_{22} - G_2G_1^{-1}A_{12})G_2G_1^{-1} & A_{22} - G_2G_1^{-1}A_{12} \end{bmatrix}$$

$$NP = \begin{bmatrix} -N_{12}G_2G_1^{-1} & N_{12} \\ -N_{22}G_2G_1^{-1} & N_{22} \end{bmatrix}$$

Equality (39) thus becomes imposes the following relations:

$$\begin{aligned} N_{12} &= 0 \\ N_{22} &= A_{22} - G_2G_1^{-1}A_{12} \end{aligned} \tag{40}$$

We note that the matrix N_{12} and N_{22} are imposed by the relations (40). It follows that a necessary condition for the matrix N to be Hurwitz is that the matrix $A_{22} - G_2G_1^{-1}A_{12}$ is also Hurwitz. Thus, observer synthesis is only possible if this matrix is Hurwitz. In this case, only m poles of the observer can be chosen arbitrarily through the choice of the matrix N_{11} (we can take $N_{21} = 0$). The other poles of the observer are equal to the eigenvalues of the matrix $A_{22} - G_2G_1^{-1}A_{12}$.

In what follows we will consider a class of nonlinear systems and we will propose some sufficient conditions that allow either to simply estimate all the states of the system without any knowledge about the unknown inputs, or to jointly estimate all the states of the system and all the unknown inputs under the additional assumption that the first derivative with respect to time of each unknown input is bounded.

- **Class of non-linear systems considered**:

Let be the following class of multi-input/multi-output nonlinear systems:

$$\begin{cases} \dot{x} = f(u, x) + G(u, s)v \\ y = Cx = x^1 \end{cases} \tag{41}$$

where the state of the system$x \in \mathbb{R}^n$, $x^1 \in \mathbb{R}^p$ is the output of the system, $X \in \mathbb{R}^{n-p}$ is the part of x containing all unmeasured states; the known input $u(t) \in$ U the set of absolutely continuous functions 'with bounded derivatives from $\mathbb{R}^+$ into U a compact of $\mathbb{R}^v$;v $\in \mathbb{R}^m$ is the unknown input with $m \leq p$; $f(u, x) = \begin{pmatrix} f^1(u, x) \\ f_X(u, x) \end{pmatrix} \in \mathbb{R}^n$, $f^1(u, x) \in \mathbb{R}^p$, $f_X(u, x) \in \mathbb{R}^{n-p} and G(u, s) = \begin{pmatrix} G^1(u, s) \\ G_X(u, s) \end{pmatrix}$ is a matrix of dimension $n \times m$ where $G^1(u, s) and G_X(u, s)$ are respectively of dimension $p \times m and (n - p) \times m$ matrices; $s(t)$ is a known bounded signal whose first derivative with respect to time is also bounded; finally$\overline{C} = \begin{bmatrix} I_p & 0_{p\times(n-p)} \end{bmatrix}$.

- **Observer synthesis procedure**:

For observer synthesis, we adopt the following assumptions:

(H1) The matrix $G^1(u, s(t))$ is full rank in columns for all $u \in$ U and for all $t \geq 0$.

(H2) The derivative with respect to time of the unknown input $v(t)$ is a completely unknown function,$\varepsilon(t)$, which is uniformly bounded, i.e., $\sup_{t\geq 0}\|\varepsilon(t)\| \leq \beta_\varepsilon$ where $\beta_\varepsilon > 0$ is a strictly positive unknown real.

7 Conclusion

In this chapter, the notion of observability and observer for linear systems is presented, also the notion of sliding mode is introduced. how to design an observer for a linear dynamical system under the influence of unknown inputs is shown. The determination of the gain of the observer to guarantee its convergence leads to the resolution of a problem of the LMI type (bilinear matrix inequalities). The resolution of these LMI constraints is performed by the method based on a change of variables and which allows to determine the matrices describing the observer.

References

1. Johson C.D.: On observers for linear systems with unknown and inaccessible inputs. **1**, 825–831 (1975)
2. Muller, M.H.A.P.C.: Design of observers for linear systems with unknown inputs. **1**, 871–875 (1992)
3. Darouach, M.Z.A.S.X.M.: Full-order observer for linear systems with unknown inputs. 1, **39**(3), 606–609 (1994)
4. Wilde, F.Y.A.R.W.: Observers for linear systems with unknown inputs. 1 sur 2AC-33, (1), 677–681 (1988)
5. Guan, E.M.S.Y.: A novel approach to the design of unknown input observers. **36**(5), 1, 632–635 (1991)
6. Tu, M.C.A.J.: State and input estimation for a class of uncertain systems. **34**(6), 1757–764 (1998)
7. Saif, Y.X.A.M.: Unknown disturbance inputs estimation based on a state functional observer design. **39**(1), 1389–1398 (2003)
8. Meditch, G.H.H.J.S.: Observers for systems with unknown and inaccessible inputs. **19**(1), 637–640 (1971)
9. Dassanake, G.L.B.J.B.S.K.: Using unknown input observers to detect and isolate sensor faults in a turbofan engine. **7**(1), 6E51–6E57 (2000)
10. Wilde, F.Y.E.R.W.: Observers for linear systems with unknown inputs. **33**(1), 677 (1988)
11. Koenig, S.M.D.: Design of a class of reduced order unknown inputs nonlinear observer for fault diagnosis. **3**(1), 2143–2147 (2001)
12. Saif, Y.G.M.: A new approach to robust fault detection and identification. **29**(3), (1), 685–695 (1993)
13. O'Reilly, R.B.J.: Observer for linear system. (1), 140 (1983)
14. Borne, G.D.-T.J.P.R.F.R.E.I.Z.P.: Modélisation et identification des processus (1992)
15. Luenberger, D.: An introduction to observers. **16**(16), 596–602 (1971)
16. Kalman, R.: A new approach to linear filtering. **82**(1), 35–45 (1960)

17. Perruquetti, J.B.W.: Sliding mode control in engineering (2002)
18. Shtessel, C.E.L.A.L.Y.: Sliding mode control and observation (2014)
19. Drakunnov, V.S.: Sliding mode observer (1995)
20. Zak, B.A.S.: State observation of nonlinear uncertain dynamical systems. **32**(1), 166–170 (1987)
21. Kalsi, S.H.A.S.H.Z.K.: Unknown input and sensor fault estimation using slidingmode observers (2011)
22. Utkin, V.: Pinciple of identification using sliding regimes (1981)

Luenberger Observer of Impulsive Systems: A Survey

Boulbaba Ghanmi

Abstract In this chapter, some results on the state estimation of impulsive systems have been conducted. This problem is rarely tackled for this class of systems by researchers and they have designed an observer in the case of autonomous impulsive linear systems. Indeed, only E. A. Medina considered in [1], and [2] an observer in the case of autonomous impulsive linear systems, under the condition of strong observability.

1 Introduction

Hybrid systems, in particular, impulsive systems, have been the object of some attention in the scientific community for some years [3–5], During the last decades, an important part of the research activities in automation has been focused on the problem of state observation of nonlinear dynamical systems. This is motivated by the fact that state estimation is an important or even indispensable step in the synthesis of control laws, for the diagnosis or the supervision of industrial systems. Recently, other applications such as synchronization and decryption in communication systems have become one of the most dynamic research areas. In this context, we have conducted research on the state estimation of impulsive systems. This problem is rarely tackled for this class of systems. Indeed, only E. A. Medina considered in [1, 2] an observer in the case of autonomous impulsive linear systems, under the condition of strong observability, which we do not do in [6].

B. Ghanmi (✉)
University of Gafsa, Gafsa, Tunisia

Faculty of Sciences of Gafsa, Department of Mathematics, University campus Sidi Ahmed Zarroug, 2112 Gafsa, Tunisia

O. Naifar and A. Ben Makhlouf (eds.), *Advances in Observer Design and Observation for Nonlinear Systems*, Studies in Systems, Decision and Control 410,
https://doi.org/10.1007/978-3-030-92731-8_5

2 Observability—Observers

As for ordinary differential systems, observability and state reconstruction are two central notions in control theory.

2.1 *Observability Criteria*

The observability of a process is a very important concept. Indeed, in order to reconstruct the state of a system, it is necessary to know, a priori, if the state variables are observed or not. In particular, from an automatic point of view, the problem of observability consists in deciding if the state variables intervening in a model can be determined according to the inputs and outputs supposed to be perfectly known. We will now explain the conditions of observability in the general case of ordinary differential systems, which will be used in the construction of observers in the impulsive case later on. Let the system described by

$$\begin{cases} \dot{x} = \mathrm{f(x, u)}, \\ y = \mathrm{h(x)}, \end{cases} \tag{1}$$

with $x \in \mathbb{R}^n$ is the state vector, $u \in \mathbb{R}^m$ is the input vector, $y \in \mathbb{R}^p$ is the output vector, $f : \mathbb{R}^n \times \mathbb{R}^m \to \mathbb{R}^n$.

Definition 1 The system (1) is said to be observable on the interval $[t_0, t_f]$, if any initial condition $x(t_0) = x_0$ is uniquely determined by the input $u(t)$ and output $y(t)$ corresponding to the system for $t \in [t_0, t_f]$.

To define better this notion of observability, it is first necessary to define the notion of indiscernibility.

Definition 2 (*Indiscernibility*) Two distinct initial states x_0 and $\overline{x}_0$ are said to be indistinguishable if, for any input function $u(t)$ and for all $t \geq t_0 \geq 0$, the resulting outputs $h(x(t, x_0, t_0))$ and $h(x(t, \overline{x}_0, t_0))$ are equal.

This means that from two different initial conditions, the output of the system is the same in both cases, and this is for the same applied command. Thus

Definition 3 (*Observability*) The system (1) is said to be observable if it does not have any pair of distinct indistinguishable initial states.

There exists a very simple algebraic characterization of observability in the case of linear autonomous systems

$$\begin{cases} \dot{x}(t) = \mathrm{Ax{+}Bu}, \\ y(t) = \mathrm{Cx} \end{cases} \tag{2}$$

where A, B and C are matrices of dimensions $n \times n$, $n \times m$, $p \times n$ respectively. In the literature, there is no such simple and strong criterion as the rank condition or the Kalman observability criterion stated in the following theorem.

Theorem 1 *The system (2) is observable if and only if the observability matrix*

$$\mathcal{O} = \begin{bmatrix} C \\ CA \\ \vdots \\ CA^{n-1} \end{bmatrix}$$

has rank n. We then say that the pair (A, C) is observable.

The linear time-varying systems

$$\begin{cases} \dot{x}(t) = \text{A(t)x+B(t)u}, \\ y(t) = \text{C(t)x}. \end{cases} \tag{3}$$

with bounded matrices $A(t)$, $B(t)$, and $C(t)$ also benefit from a quite convenient criterion to check.

Theorem 2 *The system (3) is said to be completely uniformly observable if there exist $\alpha > 0$ and $t_0 > 0$ such that for all $t \geq 0$, we have*

$$\Gamma(t, t+t_0) = \int_t^{t+t_0} X^T(s,t)C^T(s)C(s)X(s,t)ds \geq \alpha.I$$

where $X(t, t_0)$ is the resolvent verifying (3).

The matrix $\Gamma(t, t+t_0)$ is called the observability Grammien of the system (3). Complete observability means that the system is observable at every instant. When the model considered is non-linear, the task becomes more difficult. Indeed, in general, the observability of this type of system is not sufficient to synthesize an observer in the following. To overcome this difficulty, we will use particular structures such as the nonlinear affine systems in the control, mono output given by the following equation

$$\Gamma : \begin{cases} \dot{x}(t) = f(x(t)) + \sum_{i=1}^{m} u_i(t)g_i(t) \\ y(t) = h(x(t)) \end{cases}$$

with $x \in \Omega \subset \mathbb{R}^n$, $u \in \mathcal{U} \subset \mathbb{R}^m$, $y \in \mathbb{R}$ where Ω is an open, relatively compact physical domain in which the state of the system evolves and we are interested in the observability problem. $\mathcal{U}$ is the set of admissible values of the input. Otherwise, $\mathcal{U}$ is the set of measurable and bounded inputs $u(t)$. In this paragraph, we seek to prove a canonical normal form of observability by means of changes of bases. To do so, let us first consider the function

$$\begin{array}{rcl}\phi : \mathbb{R}^n & \longrightarrow & \mathbb{R}^n \\ x & \longmapsto & \begin{pmatrix} h(x) \\ L_f h(x) \\ \vdots \\ L_f^{n-1} h(x) \end{pmatrix} = \begin{pmatrix} z_1 \\ z_2 \\ \vdots \\ z_n \end{pmatrix} = z \end{array}$$

where L_f^k, $k = 1, \ldots, n-1$ denotes the successive Lie derivatives with respect to the vector field f:

$$L_f(h) = \sum_{i=1}^{n} f_i(x) \frac{\partial h}{\partial x_i}(x),$$

which transforms the system Γ into

$$(\Sigma) \begin{cases} \dot{z}(t) &= Az + \overline{\varphi}(z) + u\tilde{\varphi}(z) \\ y &= Cz \end{cases}$$

with

$$A = \begin{pmatrix} 0 & 1 & 0 & \cdots\cdots & 0 \\ \vdots & 0 & 1 & 0 \cdots & \vdots \\ \vdots & \vdots & 0 & \ddots \; 0 & \vdots \\ \vdots & \ddots & \ddots & \ddots \; \ddots & \ddots \\ 0 & \cdots & \cdots & \cdots & 1 \\ 0 & \cdots & \cdots & \cdots & 0 \end{pmatrix}, \; C = \begin{pmatrix} 1 & 0 & \cdots & 0 \end{pmatrix},$$

$$\overline{\varphi}(z) = \begin{pmatrix} 0 \\ \vdots \\ 0 \\ \overline{\varphi_n}(z) \end{pmatrix} = \begin{pmatrix} 0 \\ \vdots \\ 0 \\ L_f^{n-1} h(\phi^{-1}(z)) \end{pmatrix}, \; \tilde{\varphi}(z) = \begin{pmatrix} \tilde{\varphi}_1(z) \\ \tilde{\varphi}_2(z) \\ \vdots \\ \tilde{\varphi}_n(z) \end{pmatrix} = \begin{pmatrix} L_g h(\phi^{-1}(z)) \\ L_g L_f h(\phi^{-1}(z)) \\ \vdots \\ L_g L_f^{n-1} h(\phi^{-1}(z)) \end{pmatrix}.$$

In the following, we will present a very practical result, due to Gauthier and Bornard [7], which consists in giving a necessary and sufficient condition for the uniform local observability of the system Γ.

Theorem 3 ([7]) *Suppose that the chosen function ϕ is a diffeomorphism from Ω onto $\phi(\Omega)$, that the functions ϕ, $\overline{\varphi}$, $\tilde{\varphi}$ can be extended from Ω on $\mathbb{R}^n$ by globally Lipchitzian functions (respectively to any norm)(or functions of class C^∞) and that the system (Σ) is complete for all admissible (measurable bounded) input functions with values in $\mathcal{U}$, then the system (Γ) is uniformly locally observable if and only if the function $\tilde{\varphi}$ is of the form :*

$$\begin{cases} \tilde{\varphi}_1(z) = \quad \tilde{\varphi}_1(z_1) \\ \tilde{\varphi}_2(z) = \quad \tilde{\varphi}_2(z_1, z_2) \\ \quad \vdots \\ \tilde{\varphi}_n(z) = \tilde{\varphi}_n(z_1, \dots, z_n) \end{cases}$$

By analogy, the notion of observability is generalized to the case of impulsive systems, in the sense of Definition 1. Recently, for this class of systems, few criteria are provided in the general case, except some results for the linear time-varying and time-invariant case. For a detailed study of this concept in the impulsive case, the interested reader is referred to the works [1, 8–11]. For example, the result is due to Medina. Medina and Lawrence [1, 9], consists in characterizing geometrically reachable and unobservable sets in terms of invariant subspaces and in providing algorithms for their construction. With the same geometrical approach in 2005, Xie and Wang [10] provided criteria for controllability and observability. In 2002, Guan et al. [8] established a necessary and sufficient condition of observability of the non-autonomous linear system described by

$$\begin{aligned} \dot{x}(t) &= A(t)x(t) + B(t)u(t), \; t \in [t_{k-1}, t_k), \\ \Delta x(t_k) &= \qquad D_k x(t_k^-), \qquad\quad k \in \mathbb{N} \\ y(t) &= C(t)x(t) + D(t)u(t), \; t \in [t_{k-1}, t_k), \end{aligned} \tag{4}$$

where $A(t)$, $B(t)$, $C(t)$, $D(t)$ are continuous matrices of appropriate dimensions and $\forall k \geq 1$, $D_k = d_k.I$, $d_k \in \mathbb{R}$. The result is stated in the following theorem

Theorem 4 ([8]) *Suppose that* $\forall\, k \geq 1$, $1 + d_k \geq 0$. *Then the system (4) is observable on the interval* $[t_0, t_f]$, $(t_f \in (t_k, t_{k+1}])$ *if and only if the matrix*

$$M(t_0, t_f) := M(t_0, t_0, t_1) + \sum_{i=1}^{k-1} \prod_{j=1}^{i} (I + D_j) M(t_0, t_i, t_{i+1})$$

$$+ \prod_{j=1}^{k} (I + D_j) M(t_0, t_k, t_f)$$

is invertible, where

$$M(t_0, t_i, t_{i+1}) := \int_{t_i}^{t_{i+1}} X^T(s, t_0) C^T(s) C(s) X(s, t_0) ds,$$
$$i = 0, \dots, k-1$$
$$M(t_0, t_k, t_f) := \int_{t_k}^{t_f} X^T(s, t_0) C^T(s) C(s) X(s, t_0) ds$$

From this theorem, a criterion of Kalman type in the case of an autonomous system is given in the next corollary

Corollary 1 ([8]) *Under the same conditions of Theorem 4 with $A(t) = A$ and $C(t) = C$ constant matrices, the system (4) is observable if and only if the rank condition is verified.*

2.2 Definition and Role of an Observer

In the field of command, diagnosis, and monitoring, full or partial knowledge of the state at any given time is necessary to achieve such objectives. In practice, this requirement is difficult to meet, as an online measurement of these variables is often very expensive and sometimes impossible. Therefore, as soon as a control strategy requires the use of unmeasured state variables, it is essential to build an observer. The latter is an auxiliary dynamic system whose inputs consist of the input vector $u(t)$ and the output vector $y(t)$ of a system. The role of the observer or state estimator is to give an estimate $\hat{x}(t)$ of the state vector $x(t)$.The state reconstructor must have certain properties. In particular, the estimation error denoted $e = x - \hat{x}$ must tend to 0 as time tends to infinity, otherwise, $\lim_{t \to +\infty} \|e(t)\| = 0$. Generally, it is desired that, if the observer is initialized to the same initial conditions as the system to be observed ($\hat{x}(0) = x(0)$) then for any $t \geq 0$, $\hat{x}(t) = x(t)$. Finally, the synthesis of an observer thus remains a difficult issue although very important in practice.

3 The Conception of Observers for Impulsive Systems

3.1 Different Types of Continuous Observers: State of the Art

Initially, the systems addressed were linear systems, for which the Kalman and Luenberger observers gave good results. The Kalman filter is used in the case of stochastic systems by minimizing the covariance matrix of the estimation error, and the Luenberger observer has been used for deterministic linear systems. In the case of nonlinear systems, state observation is a bit trickier and there is currently no universal method for the synthesis of observers.The possible approaches are either an extension of linear algorithms or specific nonlinear algorithms. In the first case, the extension is based on a linearization of the model around an operating point. For the case of specific nonlinear algorithms, the numerous researches carried out on this subject (see [12, 13]) have given birth to many observation algorithms.

1. Non-linear transformation methods: This technique uses a change of coordinates to transform a non-linear system into a linear system. Once such a transformation is done, the use of a Luenberger type observer will suffice to estimate the state of

the transformed system, and thus the state of the original system using the inverse coordinate change.

2. Extended observers: In this case, the calculation of the gain of the observer is done from the linearized model around an operating point. This is for example the case of the extended Kalman filter and the extended Luenberger observer.
3. High gain observers: This type of observer is generally used for Lipschitzian systems. Its name is due to the fact that the gain of the chosen observer is large enough to compensate for the non-linearity of the system.
4. Generalized Luenberger Observers (GLO): This is a new type of observer that has been recently proposed for the class of autonomous systems. This new design consists of adding to the Luenberger observer a second gain inside the nonlinear part of the system.
5. Observers based on the contraction theory: This type of observer, as its name indicates, is based on the contraction theory used as a tool for convergence analysis. This technique leads to new synthesis conditions different from those provided by the previous techniques.

3.2 Luenberger-Type Observer of Impulsive Systems

The objective of this section is to extend the results of the construction of observers for usual systems, to a more general class, which is none other than impulsive systems. These observers are based on the outputs available in continuous time. Let us consider in this section the following system:

$$\begin{array}{rcll} \dot{x}(t) & = & Ax(t) + Bu(t), & t \in [t_{k-1}, t_k), \\ \Delta x(t_k) & = & D_k x(t_k^-), & \\ y(t) & = & Cx(t), & t \in t \neq t_k, \end{array} \tag{5}$$

where $x \in \mathbb{R}^n$ is the state, $u \in \mathbb{R}^m$ is the control, $y \in \mathbb{R}^p$ is the output, and the matrices A, B, D_k, and C have appropriate dimensions such that $\|D_k\| \leq d, \forall k \geq 1$. We assume that the sequence of impulses $t_0, t_1, \ldots$ is increasing and that there exists a constant $\tau > 0$ such that $t_{k+1} - t_k > \tau$. Furthermore, we assume that the pairs (A, B) and (A^T, C^T) verify the Kalman condition already stated in Theorem 2.4. In this section, we are interested in the construction of a Luenberger-type exponential observer, which we denote $\hat{x}$, i.e., a dynamical system is driven by the observations y and state $\hat{x}$ such that $x(t) - \hat{x}(t)$ tends to 0 when t tends to infinity. The idea is to copy the dynamics of the observed system and to add a term taking into account the difference between the prediction and the reality. We first establish a key lemma that will be useful in the construction of the observer.

Lemma 1 *[14] Assume that the pair (A, C) is observable. Then, for all $\lambda > 0$, there exists a gain matrix L such that the exponential of the matrix $\tilde{A} = A - LC$ verifies the inequality*

$$\|e^{t\tilde{A}}\| \leq \mu e^{-\lambda t} \quad \forall t \geq 0 \tag{6}$$

with $\mu e^{-\frac{\lambda\tau}{2}} d \leq 1$.

Remark 1 [14] It should be noted that it is the choice of the eigenvalues λ_i that conditions the stability and convergence performances of the observer, which we construct in the following paragraph.

In this sense and analogously to the continuous observer, we consider a Luenberger observer of the following form:

$$\begin{aligned} \dot{\hat{x}}(t) &= A\hat{x}(t) + Bu(t) + LC(x(t) - \hat{x}(t)),\ t \in [t_{k-1}, t_k), \\ \hat{x}(t_k) &= D_k\hat{x}(t_k^-). \end{aligned} \tag{7}$$

Thus the error equation is :

$$\begin{aligned} \dot{e}(t) &= (A - LC)e(t),\ t \in [t_{k-1}, t_k), \\ \Delta e(t_k) &= D_k e(t_k^-). \end{aligned} \tag{8}$$

Theorem 5 *[14] Suppose the pair* (A, C) *is observable and there exist constants* $d > 0$, $\tau > 0$ *such that* $\|D_k\| \leq d$, $t_{k-1} - t_k > \tau$, $\forall k \geq 1$. *Then, one can choose the gain matrix L such that the system (8) is globally exponentially stable. Moreover, the speed of convergence can be arbitrarily large depending on the choice of the matrix L.*

3.2.1 Stabilization and Separation Principle

The origin of the term "Separation Principle" comes from the study of the stability of autonomous linear systems commanded by estimating state feedback. Let a process be modeled in the state formalism by ordinary differential equations. Let be a state feedback control that stabilizes the considered process. Because of technical reasons, costs,..., the knowledge of the whole state is not available. Therefore, it is not possible to compute this stabilizing command law. A way out lies in the construction of an estimate of the state variables. The fundamental question is: does the command calculated from the estimates still stabilize the looped system? For non-linear systems, this is generally not the case.

Stabilization: Consider the impulsive linear system

$$\begin{aligned} \dot{x}(t) &= Ax(t) + Bu(t), \quad t \neq t_k, \\ x(t_k) &= D_k x(t_k^-), \quad k \in \mathbb{N}, \\ Y(t) &= Cx(t) + Du(t),\ t \in [t_{k-1}, t_k), \end{aligned} \tag{9}$$

The objective of this section is to construct a linear control law $u = Kx$ that exponentially stabilizes the closed-loop system

$$\begin{cases} \dot{x} &= (A - BK)x \\ x(t_k) &= D_k x(t_k^-) \end{cases} \tag{10}$$

Theorem 6 *[14] If we assume that the pair (A, B) is controllable and there exists a constant $d > 0$ such that $\|D_k\| \leq d$, $\forall k \geq 1$, then we can choose a gain matrix K such that the closed-loop system (10) is globally exponentially stable. Moreover, the speed of convergence can be arbitrarily large as one wants, depending on the choice of matrix K.*

Separation principle: We have seen previously that it is possible, under certain natural assumptions, to construct a stabilization law and an observer for the system (5), in this section we will show a separation principle. Let K be a gain matrix such that the system (10) is exponentially stable and L a gain matrix chosen such that the system (7) is an exponential observer of (5). We consider the following system obtained by joining the system (10) and (5).

$$\begin{cases} \dot{x} &= Ax - BK\hat{x} \\ \dot{\hat{x}} &= A\hat{x} - BK\hat{x} - L(C\hat{x} - y) \\ x(t_k) &= D_k x(t_k^-) \\ \hat{x}(t_k) &= D_k \hat{x}(t_k^-) \end{cases}$$

In order to show that the origin $(0, 0)$ of this system is exponentially stable, we can rewrite it by considering $e = \hat{x} - x$:

$$\begin{cases} \dot{x} &= (A - BK)x - BKe \\ \dot{e} &= (A - LC)e \\ x(t_k) &= D_k x(t_k^-) \\ \hat{e}(t_k) &= D_k \hat{e}(t_k^-) \end{cases} \tag{11}$$

Theorem 7 *[14] Assume that the pairs (A, B) and (A, C) are controllable and observable respectively and that there exists a constant $d > 0$ such that $\|D_k\| \leq d$, $\forall k \geq 1$. Thus, one can choose the gain matrices K and L such that the system (11) is exponentially stable.*

3.2.2 Luenberger-Type Observer for a Class of Perturbed Systems

In this section, we propose to study a family of autonomous systems which is a linear impulsive autonomous system perturbed by a continuous or piecewise continuous or piecewise continuous time-varying matrix. We claim, in fact, to construct a Luenberger observer for the system described by

$$\begin{array}{lll} \dot{x}(t) & = Ax(t) + Bu(t) + P(t)x(t), \ t \in [t_{k-1}, t_k), & \\ \Delta x(t_k) = & D_k x(t_k^-) + J_k x(t_k^-), & \\ x(t_0^+) & = \quad x_0, & t_0 \geq 0, \end{array} \tag{12}$$

with $A \in \mathbb{R}^{n\times n}$, $B \in \mathbb{R}^{n\times n}$, $P(t) \in \mathbb{R}^{n\times n}$ is a matrix that can be continuous or piecewise continuous for all $t \geq t_0$ and $J_k \in \mathbb{R}^{n\times n}$ is a constant matrix. Moreover, we have

($\mathcal{H}_1$) for rank k and time t large enough, $\|P(t)\| < \xi$, $\|J_k\| < \xi$.

($\mathcal{H}_2$) $0 < \tau_1 \leq t_k - t_{k-1} \leq \tau_2, \forall k \geq 1$.

Then let the next observer be.

$$\begin{array}{ll} \dot{\hat{x}}(t) & = A\hat{x}(t) + Bu(t) + P(t)\hat{x}(t) + LC(x(t) - \hat{x}(t)), \ t \in [t_{k-1}, t_k), \\ \Delta\hat{x}(t_k) = & D_k\hat{x}(t_k^-) + J_k\hat{x}(t_k^-), \end{array} \tag{13}$$

In fact, this observer realizes a compromise between the information on the state provided by the measurements (more or less tainted by errors) and the prediction of the state provided by the system (13). Thus the dynamics of the error is given by

$$\begin{array}{ll} \dot{e}(t) & = (A - LC)e(t) + P(t)e(t), \ t \in [t_{k-1}, t_k), \\ \Delta e(t_k) = & D_k e(t_k^-) + J_k e(t_k^-), \end{array}$$

Theorem 8 *[14] Assume that the system (12) is observable and that properties ($\mathcal{H}_1$) and ($\mathcal{H}_2$) are verified. Let the matrix L be chosen such that the inequalities of Lemma 3.1 are verified. Then the system (13) is an exponential observer for the perturbed linear system (12).*

3.3 Observers for Nonlinear Impulsive Systems

3.3.1 Extended Kalman Filter for a Class of Nonlinear Impulsive Systems Linear Systems

In this section, we focus on the case of autonomous nonlinear systems, which have already been treated by Gauthier and Kupka, but with impulses made at times $(t_k)_{k\geq 1}$ in three different ways. Assume that ($\mathcal{H}_1$) The system Γ is globally in the normal form (Ω).

($\mathcal{H}_2$) The functions φ and g_i, $1 \leq i \leq n$ are globally Lipschitzian respectively at $\underline{x_i} = (x_0, \ldots, x_i)$.

This is a nonlinear system to which we add a quantity of matter $p > 0$ at $(t_k)_{k\geq 1}$ reset times. We are interested precisely in the first instance in the system described by

$$\begin{aligned} \dot{x}(t) &= Ax(t) + b(x,u),\ t \in [t_{k-1}, t_k), \\ x(t_k^+) &= x(t_k^-) + p, \\ y(t) &= Cx(t), \qquad t \in [t_{k-1}, t_k), \end{aligned} \tag{14}$$

with $A = \begin{pmatrix} 0 & 1 & 0 & \cdots\cdots & 0 \\ \vdots & 0 & 1 & 0 \ \cdots & \vdots \\ \vdots & \vdots & 0 & \ddots \ 0 & \vdots \\ \vdots & \ddots & \ddots & \ddots \ \ddots & \ddots \\ 0 & \cdots & \cdots & \cdots & 1 \\ 0 & \cdots & \cdots & \cdots & \varphi(x) \end{pmatrix}$, $b(x,u) = u \begin{pmatrix} g_1(x_1) \\ g_1(x_1, x_2) \\ \vdots \\ \vdots \\ g_{n-1}(x_1, \ldots, x_{n-1}) \\ g_n(x_n) \end{pmatrix}$, $C = \begin{pmatrix} 1 & 0 & \cdots & 0 \end{pmatrix}$ and $p \in \mathbb{R}^+$.

Let now, $Q \in \mathbb{R}^{n\times n}$ be a symmetric positive definite matrix, r, θ positive reals. So lets construct the matrices $\Delta_\theta = diag(1, \frac{1}{\theta}, ..., (\frac{1}{\theta})^{n-1})$ and $Q_\theta = \theta^2(\Delta_\theta)^{-1}Q(\Delta_\theta)^{-1}$. Let the extended Kalman filter be given by

$$\begin{aligned} \dot{z}(t) &= Az(t) + b(z,u) - S^{-1}(t)C^T r^{-1}(Cz - y(t)),\ t \in [t_{k-1}, t_k), \\ z(t_k^+) &= z(t_k^-) + p, \\ \frac{dS(t)}{dt} &= -A^T S - SA + C^T r^{-1} C - SQ_\theta S. \end{aligned} \tag{15}$$

Denote $\varepsilon(t) = z(t) - x(t)$. The error system is continuous and given by :

$$\begin{aligned} \dot{\varepsilon}(t) &= A\varepsilon(t) + (b(\varepsilon + z, u) - b(z,u)) - S^{-1}(t)C^T r^{-1}(Cz - y(t)),\ t \in [t_{k-1}, t_k), \\ \varepsilon(t_k) &= \varepsilon(t_k^-). \end{aligned} \tag{16}$$

Theorem 9 *[14] Under assumptions $(\mathcal{H}_1)$ and $(\mathcal{H}_2)$, for some real $\theta > 1$ and for all $t_0 > 0$, the extended Kalman filter (15), verifies $\forall t \geq \frac{t_0}{\theta}$*

$$\|\varepsilon(t)\| \leq \theta^{n-1} k(t_0) \|\varepsilon(\frac{t_0}{\theta})\| e^{-(\theta\omega(t_0) - \mu(t_0))(t - \frac{t_0}{\theta})}$$

for continuous functions $k(t_0)$, $\omega(t_0)$, $\mu(t_0)$.

If we place in a more general case of systems that are described by

$$\begin{aligned} \dot{x}(t) &= Ax(t) + b(x,u),\ t \in [t_{k-1}, t_k), \\ \Delta x(t_k) &= s(x(t_k^-)), \\ y(t) &= Cx(t) \qquad t \in [t_{k-1}, t_k), \end{aligned} \tag{17}$$

where the matrices A, C, and the function b are defined as before (in the first case), as well as the function $s(.)$ is assumed globally Lipchitzian of Lipchitz constant equal to λ. Let then be the following extended Kalman filter

$$\begin{array}{lcl} \dot{\hat{x}}(t) & = & A\hat{x}(t) + b(\hat{x}, u) - S^{-1}(t)C^T r^{-1}(C\hat{x} - y(t)),\ t \in [t_{k-1}, t_k), \\ \Delta\hat{x}(t_k) & = & s(\hat{x}(t_k^-)), \\ \frac{dS(t)}{dt} & = & -A^T S - SA + C^T r^{-1} C - SQ_\theta S. \end{array} \tag{18}$$

Write $\varepsilon(t) = x(t) - \hat{x}(t)$. This gives the following error equation

$$\begin{array}{lcl} \dot{\varepsilon}(t) & = & A\varepsilon(t) + (b(\varepsilon + \hat{x}, u) - b(\hat{x}, u)) + S^{-1}(t)C^T r^{-1}(C\hat{x} - y(t)),\ t \in [t_{k-1}, t_k), \\ \Delta\varepsilon(t_k) & = & s(\hat{x}(t_k) + \varepsilon(t_k)) - s(\hat{x}(t_k)) \end{array} \tag{19}$$

Theorem 10 *[14] By assumptions* $(\mathcal{H}_1)$ *and* $(\mathcal{H}_2)$, *for some real* $\theta > 1$ *and for all* $t_0 > 0$, *the extended Kalman filter (18) is an exponential observer for the system (17). Moreover, it verifies* $\forall\ t \geq t_0$

$$\|\varepsilon(\frac{t}{\theta})\| \leq \frac{\beta(t_0)}{\alpha(t_0)} \sqrt{\prod_{t_0 < t_k < t} (1+\lambda)^2 \|\varepsilon(\frac{t}{\theta})\| \exp^{\frac{-1}{2}\left(Q_m \alpha(t_0) - \frac{\beta(t_0)}{\alpha(t_0)} \frac{L}{\theta}\right)} (t - t_0)},\ \forall\ t \geq t_0.$$

Now consider the system described by

$$\begin{array}{lcl} \dot{x}(t) & = & Ax(t) + b(x, u),\ t \in [t_{k-1}, t_k), \\ \Delta x(t_k^+) & = & p_k x(t_k^-), \\ y(t) & = & Cx(t) \qquad t \in [t_{k-1}, t_k), \end{array} \tag{20}$$

where the matrices A, C and the function b are denoted as before (in the first case), as well as $p \geq p_k \geq 0$. The extended Kalman filter is the following

$$\begin{array}{lcl} \dot{\hat{x}}(t) & = & A\hat{x}(t) + b(\hat{x}, u) - S^{-1}(t)C^T r^{-1}(C\hat{x} - y(t)),\ t \in [t_{k-1}, t_k), \\ \Delta\hat{x}(t_k^+) & = & \hat{x}(t_k^-) + p_k\hat{x}(t_k^-), \\ \frac{dS(t)}{dt} & = & -A^T S - SA + C^T r^{-1} C - SQ_\theta S. \end{array} \tag{21}$$

Write $\varepsilon(t) = \hat{x}(t) - x(t)$. This gives the following error equation

$$\begin{array}{lcl} \dot{\varepsilon}(t) & = & A\varepsilon(t) + (b(\varepsilon + \hat{x}, u) - b(\hat{x}, u)) + S^{-1}(t)C^T r^{-1}(C\hat{x} - y(t)),\ t \in [t_{k-1}, t_k), \\ \Delta\varepsilon(t_k^+) & = & \hat{\varepsilon}(t_k^-) + p_k\hat{\varepsilon}(t_k^-) \end{array} \tag{22}$$

Corollary 2 *[14] By assumptions* $(\mathcal{H}_1)$ *and* $(\mathcal{H}_2)$, *for some real* $\theta > 1$ *and for all* $t_0 > 0$, *the extended Kalman filter (21) is an exponential observer for the system (20). Moreover, it verifies* $\forall\ t \geq t_0$

$$\|\varepsilon(\frac{t}{\theta})\| \leq \sqrt{\frac{\beta(t_0)}{\alpha(t_0)}} \prod_{t_0 < t_k < t} (1+p) \|\varepsilon(\frac{t}{\theta})\| \exp^{\frac{-1}{2}\left(Q_m \alpha(t_0) - \frac{\beta(t_0)}{\alpha(t_0)} \frac{L}{\theta}\right)} (t - t_0),\ \forall\ t \geq t_0.$$

3.3.2 Another Type of Observer for a Class of Nonlinear Impulsive Systems

In this section, we will give other observers for the same classes of systems studied in the previous section. First, we focus on the system described by

$$\begin{array}{rcl} \dot{x}(t) & = & Ax(t)+b(x,u),\ t\in[t_{k-1},t_k),\\ x(t_k^+) & = & x(t_k^-)+p,\\ y(t) & = & Cx(t),\qquad t\in[t_{k-1},t_k), \end{array} \tag{23}$$

with $A=\begin{pmatrix} 0 & 1 & 0 & \cdots\cdots & 0\\ \vdots & 0 & 1 & 0 \ \cdots & \vdots\\ \vdots & \vdots & 0 & \ddots \ 0 & \vdots\\ \vdots & \ddots & \ddots & \ddots \ \ddots & \ddots\\ 0 & \cdots & \cdots & \cdots & 1\\ 0 & \cdots & \cdots & \cdots & \varphi(x) \end{pmatrix}$, $b(x,u)=u\begin{pmatrix} g_1(x_1)\\ g_1(x_1,x_2)\\ \vdots\\ \vdots\\ g_{n-1}(x_1,\ldots,x_{n-1})\\ g_n(x_n) \end{pmatrix}$, $C=\begin{pmatrix}1\ 0\cdots 0\end{pmatrix}$ and $p\in\mathbb{R}^+$. Suppose further that assumptions $(\mathcal{H}_1)$, $(\mathcal{H}_2)$ are satisfied,

$(\mathcal{H}_1)$ The system is globally in normal form (Σ).

$(\mathcal{H}_2)$ The functions φ and g_i, $1\le i\le n$ are globally Lipschitzian respectively at $\underline{x_i}=(x_0,\ldots,x_i)$.

Let $S_t(\theta)\in\Sigma^+$: be the cone of positive definite symmetric matrices, such that

$$\dot{S}_t(\theta)=-\theta S_t(\theta)-A'S_t(\theta)-S_t(\theta)A+C'C$$

and let $S_\infty(\theta)=\lim\limits_{t\to\infty}S_t(\theta)$. Let then be the dynamic system given by

$$\begin{array}{rcl} \dot{z}(t) & = & Az(t)+b(z,u)-S_\infty(\theta)^{-1}(t)C'r^{-1}(Cz-y(t)),\ t\in[t_{k-1},t_k),\\ z(t_k^+) & = & z(t_k^-)+p,\\ 0 & = & -\theta S_\infty(\theta)-A'S_\infty(\theta)-S_\infty(\theta)A+C'C. \end{array} \tag{24}$$

Write $\varepsilon(t)=z(t)-x(t)$. The error system is continuous and is given by:

$$\begin{array}{rcl} \dot{\varepsilon}(t) & = & A\varepsilon(t)+(b(\varepsilon+z,u)-b(z,u))+S_\infty(\theta)^{-1}(t)C^Tr^{-1}(Cz-y(t)),\ t\in[t_{k-1},t_k),\\ \varepsilon(t_k^+) & = & \varepsilon(t_k^-). \end{array} \tag{25}$$

Theorem 11 *[14] Under assumptions $(\mathcal{H}_1)$ and $(\mathcal{H}_2)$, for uniformly bounded inputs u by u_0, $t_0>0$, system (24) is an exponential observer for system (23), and for θ sufficiently large, we have $\forall t\ge t_0$*

$$\|\varepsilon(t)\| \leq \sqrt{\frac{\lambda_{\max}(S_\infty(\theta))}{\lambda_{\min}(S_\infty(\theta))}}\|\varepsilon(t_0)\|e^{-\frac{1}{2}(\theta-2kn\sqrt{C_1S})(t-t_0)}$$

Now let's place ourselves in a more general case of systems that are described by

$$\begin{array}{rcl} \dot{x}(t) &=& Ax(t)+b(x,u),\ t\in[t_{k-1},t_k),\\ \Delta x(t_k) &=& s(x(t_k^-)),\\ y(t) &=& Cx(t) \qquad t\in[t_{k-1},t_k),\end{array} \tag{26}$$

where the matrices A, C, and the function b are defined as before (in the first case), as well as the function $s(t)$ is assumed globally Lipchitzian of Lipchitz constant equal to $\lambda > 0$. Let then be the following extended Kalman filter

$$\begin{array}{rcl} \dot{z}(t) &=& Az(t)+b(z,u)-S_\infty(\theta)^{-1}(t)C'(Cz-y(t)),\ t\in[t_{k-1},t_k),\\ \Delta z(t_k) &=& s(z(t_k^-)),\\ \frac{dS(t)}{dt} &=& -A^TS-SA+C^Tr^{-1}C-SQ_\theta S.\end{array} \tag{27}$$

Write $\varepsilon(t)=x(t)-z(t)$. This gives the following error equation

$$\begin{array}{rcl} \dot{\varepsilon}(t) &=& A\varepsilon(t)+(b(\varepsilon+z,u)-b(z,u))+S^{-1}(t)C^Tr^{-1}(Cz-y(t)),\ t\in[t_{k-1},t_k),\\ \Delta\varepsilon(t_k) &=& s(z(t_k)+\varepsilon(t_k))-s(z(t_k)) \end{array} \tag{28}$$

Theorem 12 *[14] Under assumptions ($\mathcal{H}_1$) and ($\mathcal{H}_2$), for uniformly bounded inputs u by u_0, $t_0 > 0$, system (27) is an exponential observer for system (26), and for θ sufficiently large, we have $\forall t \geq t_0$*

$$\|\varepsilon(t)\| \leq \frac{\lambda_{\max}(S_\infty(\theta))}{\lambda_{\min}(S_\infty(\theta))}\prod_{t_0<t_k<t}(1+\lambda)\|\varepsilon(t_0)\|e^{-\left(\frac{\theta-2kn\sqrt{C_1S}}{2}\right)(t-t_0)},\ \forall t\geq t_0$$

As in the previous section, we can consider a special case of systems described by

$$\begin{array}{rcl} \dot{x}(t) &=& Ax(t)+b(x,u),\ t\in[t_{k-1},t_k),\\ x(t_k^+) &=& x(t_k^-)+px(t_k^-),\\ y(t) &=& Cx(t), \qquad t\in[t_{k-1},t_k),\end{array} \tag{29}$$

where the matrices A, C, and the function b are denoted as before, as well as $p \geq 0$. Let the following observer be

$$\begin{array}{rcl} \dot{z}(t) &=& Az(t)+b(z,u)-S_\infty(\theta)^{-1}(t)C'(Cz-y(t)),\ t\in[t_{k-1},t_k),\\ z(t_k^+) &=& z(t_k^-)+pz(t_k^-),\\ 0 &=& -\theta S_\infty(\theta)-A'S_\infty(\theta)-S_\infty(\theta)A+C'C.\end{array} \tag{30}$$

Note by $\varepsilon(t)=x(t)-z(t)$. This gives the following error system:

$$\begin{aligned} \dot{\varepsilon}(t) &= A\varepsilon(t) + (b(\varepsilon + z, u) - b(z, u)) + S^{-1}(t)C^T r^{-1}(Cz - y(t)),\ t \in [t_{k-1}, t_k), \\ \varepsilon(t_k^+) &= \varepsilon(t_k^-) + p\varepsilon(t_k^-). \end{aligned} \tag{31}$$

Corollary 3 *[14] Under assumptions ($\mathcal{H}_1$) and ($\mathcal{H}_2$), for uniformly bounded inputs u by u_0, $t_0 > 0$, system (30) is an exponential observer for system (29), and for θ sufficiently large, we have $\forall t \geq t_0$*

$$\|\varepsilon(t)\| \leq \sqrt{\frac{\lambda_{\max}(S_\infty(\theta))}{\lambda_{\min}(S_\infty(\theta))}} \prod_{t_0 < t_k < t} (1 + p)\|\varepsilon(t_0)\| e^{-\frac{1}{2}(\theta - 2kn\sqrt{C_1 S})(t - t_0)}$$

References

1. Medina, E.A.: Linear impulsive control systems: a geometric approach. Ph.D. thesis, Ohio University, School of Electrical Engineering and Computer Science (2007)
2. Medina, E.A., Lawrence, D.A.: State estimation for linear impulsive systems. ACC'09. In: Proceedings of the 2009 Conference on American Control Conference (2009)
3. Liu, X.: Stability of impulsive control systems with time delay. Math. Comput. Model. **39**, 511–519 (2004)
4. Teel, A.R.: Robust hybrid control systems: an overview of some recent results. In: Lect. Notes Control Inf. Sci. **353**, 279–302 (2007)
5. Yuan, R., Jing, Z., Chen, L.: Uniform asymptotic stability of hybrid dynamical systems with delay. IEEE Trans. Autom. Control **48**(2), 344–348 (2003)
6. Ellouze, I., Vivalda, J.C., Hammami M.A.: A separation principle for linear impulsive systems. Eur J Control, **20**, 105–110 (2014)
7. Gauthier, J.P., Bornard, G.: Observability for any u(t) of a class of nonlinear systems. IEEE Trans. Autom. Control **26**(4), 922-926 (1981)
8. Guan, Z.H., Qian, T.H., Yu, X.: On controllability and observability for a class of impulsive systems. Syst. Control Lett. **47**(3), 247–257 (2002)
9. Medina, E.A., Lawrence, D.A.: Reachability and observability of linear impulsive systems. Automatica 44(5), 1304–1309 (2008)
10. Xie, G., Wang, L.: Controllability and observability of a class of linear impulsive systems. J. Math. Anal. Appl. **304**(1), 36–355 (2005)
11. Zhaoa, S., Sun, J.: Controllability and observability for a class of time-varying impulsive systems. Nonlinear Anal. Real World Appl. **10**(3), 1370–1380 (2009)
12. Misawa, E.A., Hedrick, J.K.: Nonlinear observers-a state of the art survey. J. Dyn. Syst. Meas. Control **111**, 344–352 (1989)
13. Walcott, B.L., Corless, M.J., Zak, S.K.: Comparative study of nonlinear state obser- vation techniques. Int. J. Control **45**(6), 2109–2132 (1987)
14. Imen Ellouze. Etude de la stabilité et de la stabilisation des systèmes à retard et des systèmes impulsifs. Optimisation et contrôle [math.OC]. Université Paul Verlaine - Metz; Université de Sfax, 2010. Français

Compensator Design Via the Separation Principle for a Class of Nonlinear Uncertain Evolution Equations on a Hilbert Space

Hanen Damak

Abstract The compensator design via the separation principle for evolution equations in Hilbert spaces has been the subject of several studies. Under uniformly Lipschitz continuous perturbation, we propose a non-linear time-varying Luenberger observer to estimate the system states and we prove that the Luenberger observer based on linear controller stabilizes the system. These results are applied to partial differential equations.

Keywords Compensator design · Luenberger observer · Hilbert spaces · Uncertain evolution equations · Stabilization

2000 Mathematics Subject Classification 93C10 · 93C25 · 93B15

1 Introduction

The feedback compensator design of partial differential equations has been attracting a lot of attention (see [3–5, 8, 9, 24]) over the past few decades. For linear systems this problem is completely solved (see [4, 9]), but if the system contains some non-linearities as perturbation or disturbances, the problem in observer design still remains a difficult task. The theory of compensator design is a straightforward extension of the finite dimensional theory and has been used as a starting point in many control designs for distributed parameter systems, see [17, 19, 21, 25]. Alternative direct state-space finite-dimensional compensator designs can be found in [3, 7]. For extensions to systems with unbounded input and output operators (see [5, 8]), and for a comparison of various finite-dimensional control designs, see [6]. In recent years, non-autonomous differential equations on infinite-dimensional spaces have been studied by many researchers, see [8–10, 12–15] for more details. The authors in [24] give an observer-based output feedback design for linear parabolic

H. Damak (✉)
Department of Mathematics, Faculty of Sciences of Sfax, University of Sfax, Road of Soukra BP:1171, 3000 Sfax, Tunisia

O. Naifar and A. Ben Makhlouf (eds.), *Advances in Observer Design and Observation for Nonlinear Systems*, Studies in Systems, Decision and Control 410,
https://doi.org/10.1007/978-3-030-92731-8_6

partial differential equation (PDE) with local piecewise control and pointwise observation in space. In [14], we showed that we can find an exponentially stabilizing compensator for a class of semilinear evolution equations if and only if the nominal system is exponentially stabilizable and detectable. However, the problem of stabilization of the infinite-dimension time-varying control systems in Hilbert spaces has been presented in [10, 12, 15]. In finite dimensions one simple way of designing a compensator is to first construct a state feedback stabilizer and an observer for the system and then to combine the two to design a compensator using a feedback of the observer instead of the state. This is the so-called separation principle, see [1, 2, 11, 16, 18, 20, 22].

The main contribution of this chapter is the study of the problem of feedback stabilization for a class of nonlinear uncertain evolution equations via a state controller. More specifically, we design a non-linear observer to estimate the system states. Furthermore, we prove a compensator design, that is, we use the measurements (partial information) to estimate the full state (the construction of an observer) and to apply state feedback on the estimated state.

The organization of the rest of this chapter is given as follows. The system description, notations and some preliminary results are presented in Sect. 2. The required assumptions and the statement of the main results are provided in Secti. 3. In Sect. 4, an example of application of the result is given. Finally, our conclusion is presented in Sect. 5.

2 Preliminaries

Notations. $\mathbb{R}_+$ denotes the set of all non-negative real numbers, H denotes a Hilbert space with the norm $\|.\|$ and the inner product $\langle ..\rangle$. Also,

- $L(X)$ (respectively, $L(X, Y)$) denotes the space of all linear bounded operators S mapping X into X (respectively, X into Y) endowed with the norm

$$\|S\| = \sup\{\|Sx\| : x \in X, \|x\| \leq 1\}.$$

- The domain and the adjoint of an operator A are denoted by $D(A)$ and A^* respectively. I as the identity operator.
- $C([0, \infty), H)$ denotes the space of all continuous functions from $[0, \infty)$ to H.
- $\mathbf{1}_{[\vartheta,\varsigma]} = \begin{cases} 1, \text{ si } \vartheta \leq x \leq \varsigma \\ 0, \text{ elsewhere.} \end{cases}$

Consider the infinite-dimensional controlled system of the form:

$$\begin{cases} \dot{x}(t) = Ax(t) + Bu(t) + F(t, x), \quad t \geq 0, \\ \\ y = Cx, \end{cases} \tag{1}$$

where $x \in H$ is the system state, $u \in U$ is the control input, $y \in Y$ is the measured output. H, U and Y are assumed to be Hilbert spaces. Further, the operator $A : D(A) \subset H \to H$ is the generator of a C_0-semigroup over H with a domain of definition $D(A)$, $B \in L(U, H)$, $C \in L(H, Y)$ and $F : \mathbb{R}_+ \times H \longrightarrow H$ is a nonlinear function and $F(t, 0) = 0$.

Given an initial condition

$$x(t_0) = x_0 \in H.$$

Let $x(t) = x(t, t_0, x_0, u)$ denote the state of system (1) at moment $t \geq t_0 \geq 0$ associated with an initial condition $x_0 \in X$ at $t = t_0$ and input $u \in U$.

We consider mild solutions of (1), i.e. solutions of the integral equation

$$x(t) = T(t - t_0)x_0 + \int_{t_0}^{t} T(t - s)[Bu(s) + F(s, x(s))]ds, \tag{2}$$

belonging to the class $C([t_0, t_0 + \delta], H)$ for certain $\delta > 0$. Here $\{T(t), t \geq 0\}$ is a C_0- semigroup on a Hilbert space H with an infinitesimal generator $A : D(A) \subset H \to H$, $Ax = \lim\limits_{t \longrightarrow 0^+} \dfrac{T(t)x - x}{t}$, whose domain of definition $D(A)$ consists of those $x \in H$, for which this limits exists.

We will use the following assumption concerning nonlinearity F.

($\mathcal{H}_1$) The function $F : \mathbb{R}_+ \times H \longrightarrow H$ is continuous and there exists $K(.) \in L^1(0, \infty)$, such that

$$\|F(t, x) - F(t, y)\| \leq K(t)\|x - y\|$$

holds for all $x, y \in H$.

The corresponding system without perturbations, called the nominal system, is described by

$$\dot{x}(t) = Ax(t), \quad x(0) = x_0, \quad t \geq 0. \tag{3}$$

Next, we recall the definition of the generator of an exponentially stable semi-group as well as that of the exponential stabilizability and detectability, see Curtain and Zwart [9] for details.

Definition 1 The operator A generates an exponentially stable semigroup $T(t)$ if the initial value problem (3) has a unique solution $x(t) = T(t)x_0$, and

$$\|T(t)\| \leq Me^{-\alpha t}, \quad \text{for all } t \geq 0,$$

with some positive numbers M and α.

The α is called the decay rate.

If $T(t)$ is exponentially stable, then the solution to the abstract Cauchy problem (3) tends to zero exponentially as $t \longrightarrow \infty$.

An important criterion for exponential stability is the following.

Lemma 1 *The C_0*−*semigroup $T(t)$ on H is exponentially stable if and only if for every $x \in H$, there exists a positive constant γ_x, such that*

$$\int_0^\infty \|T(t)x\|^2 dt \leq \gamma_x.$$

Definition 2 The pair $\{A, B\}$ is said to be exponentially stabilizable if there exists a feedback operator $D \in L(H, U)$, such that the operator $A + BD$ generates an exponentially stable semigroup T_{BD}.

Definition 3 The pair $\{A, C\}$ is said to be exponentially detectable if there exists an output injection operator $L \in L(Y, U)$, such that the operator $A + LC$ generates an exponentially stable semigroup T_{LC}.

To study stability properties of (1) with respect to external inputs, we use the notion of stabilizability.

Definition 4 System (1) is stabilizable if there exists a continuous feedback control $u : X \to U$, such that system (1) with $u(t) = u(x(t))$ satisfies the following properties:

(i) For any initial condition $x_0 \in H$, there exists a unique mild solution $x(t)$ defined on $\mathbb{R}_+$.
(ii) There exist positive scalars ω, k, such that the solution of the system (1) satisfies

$$\|x(t)\| \leq k\|x_0\|e^{-\omega(t-t_0)}, \quad \forall t \geq t_0 \geq 0.$$

When (i) and (ii) are satisfied for (1), we say that (1) with $u(t) = u(x(t))$ is globally uniformly exponentially stable.

Remark 1 We deal with the stabilizability of (1) whose the origin is an equilibrium point. In the case of infinite-dimensional space, the stabilizability is studied by [12] of a class of time-varying control systems having a time-varying linear part.

In what follows, we shall that $V : X \to \mathbb{R}_+$ is a Lyapunov function.

Definition 5 The Lie derivative of V corresponding to the input u is defined by

$$\dot{V}(x) = \limsup_{t \to 0^+} \frac{1}{t}(V(x(t, x, u) - V(x)).$$

Recall that a self-adjoint operator $P \in L(H)$ is called positive if $\langle Px, x\rangle > 0$ holds for all $x \in H\backslash\{0\}$. A positive operator $P \in L(H)$ is called coercive if there exists $k > 0$, such that $\langle Px, x\rangle \geq k\|x\|^2, \quad \forall x \in D(P)$.

Proposition 1 (See [9]) *Suppose that A is the infinitesimal generator of the C_0−semigroup $T(t)$ on the Hilbert space H. Then, $T(t)$ is exponentially stable if and only if there exists a coercive positive self-adjoint operator $P \in L(H)$, such that*

$$\langle Ax, Px\rangle + \langle Px, Ax\rangle = -\langle x, x\rangle, \quad \forall x \in D(A). \tag{4}$$

Equation (4) is called a Lyapunov equation.

3 Main Results

3.1 Stabilization

We consider the non-linear system (1) satisfying the following assumptions.

($\mathcal{H}_2$) The pair $\{A, B\}$ is exponentially stabilizable, there exists a constant operator $D \in L(X, U)$, such that a sufficient condition specially related to operator $A_D = A + BD$ is presented in Curtain and Zwart [9] as the following: there exists a coercive positive self-adjoint operator P_1

$$\mu I \leq P_1 \leq \|P_1\| I,$$

where $\mu > 0$, which satisfies

$$A_D^* P_1 + P_1 A_D = -I. \tag{5}$$

We start with the following result which assures the global existence and uniqueness of mild solutions of (1).

Lemma 2 *Suppose ($\mathcal{H}_1$) holds. Then, the system (1) possesses a unique mild solution $x \in C([0, \infty), H)$ for any $x_0 \in H$.*

Proof The mild solution of (1) exists and is unique, according to a existence and uniqueness theorem in [23]). Using the fact that $u(t) = Dx(t)$, we have

$$\|x(t)\| \leq M\|x_0\| + M\left(\int_{t_0}^{t} \|B\|\|D\|\|x\| + K(t)\|x\|\right) ds. \tag{6}$$

By applying Gronwall inequality (see [26], Lemma 2.7, p42) to inequality (6), any solution of this equation is uniformly bounded

$$\|x(t)\| \leq M\|x_0\| e^{M\|B\|\|D\|\delta + M\delta M_1},$$

where $M = \sup\{\|T(t-s)\| : 0 \leq t_0 \leq s \leq t \leq t+\delta\}$ and $M_1 = \sup_{t\in[t_0,t_0+\delta]} \|K(t)\|$. Then, using Theorem 1.4 in [23], we have $t_0 + \delta = \infty$, and so the corresponding $x \in C([0,\infty), H)$ is a mild solution of (1). The proof is completed. □

Next, sufficient conditions are presented to guarantee the stabilizability of a perturbed control system using Lyapunov direct method.

Theorem 1 *If assumptions* $(\mathcal{H}_1)$ *and* $(\mathcal{H}_2)$ *are fulfilled, then the system* (1) *in closed-loop with the linear feedback* $u(t) = Dx$ *is globally uniformly exponentially stable.*

Proof Let $x(t)$ be the solution of system (1). We consider the Lyapunov function:

$$V(x) = \langle P_1 x, x\rangle.$$

The Lie derivative of V in t along the solution of the system (1) in the closed-loop with the controller $u(t) = Dx$ leads to

$$\begin{aligned}\dot{V}(x) &= \langle P_1\dot{x}, x\rangle + \langle P_1 x, \dot{x}\rangle\\ &= \langle P_1[(A+BD)x + F(t,x)], x\rangle + \langle P_1 x, [(A+BD)x + F(t,x)]\rangle.\end{aligned}$$

Using $(\mathcal{H}_2)$ with the help of Cauchy-Schwartz inequality, we obtain

$$\dot{V}(x) \leq -\langle x, x\rangle + 2\|P_1\|\|F(t,x)\|\|x\|.$$

It follows from assumption $(\mathcal{H}_1)$ that

$$\dot{V}(x) \leq -\langle x, x\rangle + 2\|P_1\|K(t)\|x\|^2. \tag{7}$$

It is well known that

$$\mu\|x\|^2 \leq V(x) \leq \|P_1\|\|x\|^2. \tag{8}$$

Using (8), Eq. (7) will be:

$$\dot{V}(x) \leq -\frac{1}{\|P_1\|}V(x) + \frac{2\|P_1\|K(t)}{\mu}V(x).$$

Then,

$$V(x(t)) \leq V(x(t_0))e^{-\frac{1}{\|P_1\|}(t-t_0)}e^{\frac{2\|P_1\|}{\mu}\int_0^\infty K(s)ds}.$$

Using $(\mathcal{H}_1)$, we have

$$\|x(t)\| \leq \sqrt{\frac{\|P_1\|}{\mu}}\|x_0\|e^{-\frac{1}{2\|P_1\|}(t-t_0)}e^{\frac{\|P_1\|}{\mu}\int_0^\infty K(s)ds}.$$

Consequently, the system (1) in closed-loop with the linear feedback $u(t) = Dx$ is globally uniformly exponentially stable. This finishes the proof. □

3.2 Luenberger Observer

In this subsection, we use the measurements to estimate the full state (the construction of a Luenberger observer) and to apply state feedback on the estimated state. The Luenberger observer is a dynamical system that is expected to reconstruct the states of the system. Our objective is to design a state reconstructor for the system (1), such that the global exponential stability of the resulting error system can be guaranteed.

We shall introduce the following assumptions:

($\mathcal{H}_3$) The pair $\{A, C\}$ is exponentially detectable, there exists a constant operator $L \in L(Y, H)$, such that a sufficient condition specially related to operator $A_L = A + LC$ is presented in Curtain and Zwart [9] as the following: there exists a coercive positive self-adjoint operator P_2

$$\nu I \leq P_2 \leq \|P_2\| I,$$

where $\nu > 0$, which satisfies

$$A_L^* P_2 + P_2 A_L = -I. \tag{9}$$

To design a Luenberger observer, we shall consider the system:

$$\begin{cases} \dot{\hat{x}}(t) = A\hat{x}(t) + Bu(t) + F(t, \hat{x}) + L(\hat{y}(t) - y(t)), & t \geq 0, \\ \hat{y}(t) = C\hat{x}(t), \end{cases} \tag{10}$$

where $\hat{x}$ is the Luenberger observer with output injection $L \in L(Y, Z)$.
Define estimation error e as $e = \hat{x} - x$, which is governed by

$$\dot{e}(t) = \dot{\hat{x}}(t) - \dot{x}(t) = (A + LC)e(t) + F(t, \hat{x}) - F(t, x), \tag{11}$$

where $e_0 = \hat{x}_0 - x_0$.
The following Lemma provides sufficient conditions for the global solution of (11).

Lemma 3 *Under assumption ($\mathcal{H}_1$), the system (11) possesses a unique mild solution $e \in C([0, \infty), H)$ for any $e_0 \in H$.*

Proof It is known from Pazy [23] that for every initial state $e_0 \in H$, system (11) has a unique mild solution given by

$$e(t) = T_{LC}(t-t_0)e_0 + \int_{t_0}^{t} T_{LC}(t-s)[F(s,\hat{x}(s)) - F(s,x(s))]ds, \quad t_0 \le t \le t_0+\delta, \quad \delta > 0,$$

where T_{LC} is the C_0−semigroup of A_L.
From the above equation, we get

$$\|e(t)\| \le M\|e_0\| + M\left(\int_{t_0}^{t} K(s)e(s)ds\right), \tag{12}$$

where $M = \sup\{\|T(t-s)\| : 0 \le t_0 \le s \le t \le t+\delta\}$ on an arbitrary time interval $[t_0, t_0+\delta]$. By applying Gronwall inequality (see [26], Lemma 2.7, p. 42) to inequality (12), any solution of this equation is uniformly bounded

$$\|e(t)\| \le M\|e_0\|e^{M\delta M_1},$$

where $M_1 = \sup_{t\in[t_0,t_0+\delta]} \|K(t)\|$. Then, using Theorem 1.4 in [23], we have $t_0+\delta = \infty$, and so the system (11) has a unique mild solution e which exists for all $t \ge t_0$. The proof is completed. □

By arguing in exactly the same way as in Theorem 1, we prove that the output injection L can be chosen in such a way that system (10) is an exponential Luenberger observer for system (1).

Theorem 2 *Under assumptions $(\mathcal{H}_1)$ and $(\mathcal{H}_3)$, the system (10) is an exponential Luenberger observer for the system (1).*

Proof Let $e(t)$ be the solution of system (11). We consider the Lyapunov function candidate:

$$Z(e) = \langle P_2e, e\rangle.$$

The Lie derivative of Z along any trajectory of the error Eq. (11) is given by

$$\begin{aligned}\dot{Z}(e) &= \langle P_2\dot{e}, e\rangle + \langle P_2e, \dot{e}\rangle \\ &= \langle P_2[(A+LC)e + F(t,\hat{x}) - F(t,x)], e\rangle \\ &+ \langle P_2e[(A+LC)e + F(t,\hat{x}) - F(t,x)]\rangle.\end{aligned}$$

Using $(\mathcal{H}_3)$ with the help of Cauchy-Schwartz inequality, we have

$$\dot{Z}(e) \le -\langle e, e\rangle + 2\|P_2\|\|F(t,\hat{x}) - F(t,x)\|\|e\|$$

It follows from assumption $(\mathcal{H}_1)$ that

$$\dot{Z}(e) \le -\frac{1}{\|P_2\|}Z(e) + 2\|P_2\|^2K(t)Z(e).$$

Then,

$$Z(e(t)) \leq Z(e(t_0))e^{-\frac{1}{\|P_2\|}(t-t_0)}e^{2\|P_2\|^2\int_0^\infty K(s)ds}.$$

By taking into account assumption $(\mathcal{H}_3)$, the above expression yields

$$\|e(t)\| \leq \sqrt{\frac{\|P_2\|}{\nu}}\|e_0\|e^{-\frac{1}{2\|P_2\|}(t-t_0)}e^{\|P_2\|^2\int_0^\infty K(s)ds}.$$

Thus, the system (11) is globally uniformly exponentially stable. Consequently, the system (10) is a global uniform exponential Luenberger observer for the system (1). This finishes The proof. □

3.3 Compensator Design

Here, we investigate the compensator design problem of (1). We consider the system (1) controlled by the linear feedback control $u(t) = D\hat{x}(t)$ estimated by the Luenberger observer (10).

Theorem 3 *Consider the non-linear system* (1) *and suppose that assumptions* $(\mathcal{H}_1)$, $(\mathcal{H}_2)$ *and* $(\mathcal{H}_3)$ *hold. If* $D \in L(H, U)$ *and* $L \in L(Y, U)$ *are such that* $A + BD$ *and* $A + LC$ *generate exponentially stable semigroups, then the controller* $u = D\hat{x}$, *where* $\hat{x}$ *is the Luenberger observer with output injection* L, *stabilizes the closed-loop system. The stabilizing compensator is given by*

$$\begin{cases} \dot{\hat{x}} = (A + LC)\hat{x} + Bu(t) + F(t, \hat{x}) - Ly(t) \\ u(t) = D\hat{x}(t) \end{cases} \tag{13}$$

Proof Under assumptions $(\mathcal{H}_2)$ and $(\mathcal{H}_3)$, there exist operators D and L, such that $T_{BD}(t)$ and $T_{LC}(t)$ are exponentially stable. Combining the abstract differential equations, we see that the closed-loop system is given by the dynamics of the extended state $x^e = \begin{pmatrix} \hat{x} \\ e \end{pmatrix}$,

$$\begin{pmatrix} \dot{\hat{x}} \\ \dot{e} \end{pmatrix}(t) = \begin{pmatrix} A + BD & LC \\ 0 & A + LC \end{pmatrix} \times \begin{pmatrix} \hat{x} \\ e \end{pmatrix}(t) + \begin{pmatrix} F(t, \hat{x}) \\ F(t, \hat{x}) - F(t, \hat{x} - e) \end{pmatrix}, \quad t \geq 0. \tag{14}$$

Consider the following Lyapunov function:

$$Y(x^e) = \mu V(\hat{x}) + Z(e),$$

where $V(\hat{x}) = \langle P_1\hat{x}, \hat{x}\rangle$, $Z(e) = \langle P_2 e, e\rangle$ and $\mu > 0$ is a Lyapunov parameter to be determined. Then, the Lie derivative of Y along the trajectories of system (14) is given by

$$\begin{aligned}\dot{Y}(x^e) &= \alpha\dot{V}(\hat{x}) + \dot{Z}(e)\\ &= \alpha(\langle P_1\dot{\hat{x}}, \hat{x}\rangle + \langle P_1\hat{x}, \dot{\hat{x}}\rangle) + \langle P_2\dot{e}, e\rangle + \langle P_2 e, \dot{e}\rangle\\ &= \alpha(\langle P_1[A\hat{x} + BD\hat{x} + F(t,\hat{x}) + LCe], \hat{x}\rangle + \langle P_1\hat{x}, A\hat{x} + BD\hat{x} + F(t,\hat{x}) + LCe\rangle\\ &+ \langle P_2[(A+LC)e + F(t,\hat{x}) - F(t,x)], e\rangle + \langle P_2 e, (A+LC)e + F(t,\hat{x}) - F(t,x)\rangle\end{aligned}$$

Using $(\mathcal{H}_2)$ and $(\mathcal{H}_3)$, with the help of Cauchy-Schwartz inequality, we obtain

$$\begin{aligned}\dot{Y}(x^e) \le\ & \alpha(-\langle\hat{x},\hat{x}\rangle + 2\|P_1\|\|F(t,\hat{x})\|\|\hat{x}\|\\ &+ 2\|P_1\|\|LCe\|\|\hat{x}\|) - \langle e, e\rangle + 2\|P_2\|\|F(t,\hat{x}) - F(t,x)\|\|e\|.\end{aligned}$$

It follows that,

$$\begin{aligned}\dot{Y}(x^e) \le\ & \alpha\left(-\frac{1}{\|P_1\|}V(\hat{x}) + 2\|P_1\|K(t)\|\hat{x}\|^2 + 2\|P_1\|\|LCe\|\|\hat{x}\|\right)\\ &- \frac{1}{\|P_2\|}Z(e) + 2\|P_2\|K(t)\|e(t)\|^2.\end{aligned}$$

Let $\varepsilon > 0$. Using Young's inequality

$$2\|\hat{x}\|\|e\| \le \frac{1}{\varepsilon}\|\hat{x}\|^2 + \varepsilon\|e\|^2,$$

we can continue the above estimates as

$$\begin{aligned}\dot{Y}(x^e) \le\ & \left(-\frac{1}{\|P_1\|} + \frac{2\|P_1\|K(t)}{\mu} + \frac{\|P_1\|\|LC\|}{\mu\varepsilon}\right)\alpha V(\hat{x})\\ &+ \left(-\frac{1}{\|P_2\|} + \frac{2\|P_2\|K(t)}{\nu} + \frac{\alpha\varepsilon\|P_1\|\|LC\|}{\nu}\right)Z(e).\end{aligned}$$

Choose $\varepsilon > 0$, such that $\frac{1}{\|P_1\|} - \frac{\|P_1\|\|LC\|}{\mu\varepsilon} > 0$.
Let,

$$\varepsilon = \frac{2\|P_1\|^2\|LC\|}{\mu}.$$

Also, choose for this value of ε the scalar α, such that $\frac{1}{\|P_2\|} - \frac{\alpha\varepsilon\|P_1\|\|LC\|}{\nu} > 0$. Then, let

$$\alpha = \frac{\mu\nu}{4\|P_1\|^3\|P_2\|\|LC\|^2}.$$

We get,

$$\dot{Y}(x^e) \leq \left(-\frac{\alpha}{2\|P_1\|} + \frac{2\alpha\|P_1\|K(t)}{\mu}\right)V(\hat{x}) + \left(-\frac{1}{2\|P_2\|} + \frac{2\|P_2\|K(t)}{\nu}\right)Z(e).$$

Thus,

$$\dot{Y}(x^e) \leq -\lambda_1 Y(x^e) + \lambda_2 K(t) Y(x^e),$$

with

$$\lambda_1 = \min\left(\frac{1}{2\|P_1\|}, \frac{1}{2\|P_2\|}\right),$$

and

$$\lambda_2 = \max\left(\frac{2\|P_1\|}{\mu}, \frac{2\|P_2\|}{\nu}\right).$$

Then, one have the following estimate

$$Y(x^e(t)) \leq Y(x_0^e)e^{-\lambda_1(t-t_0)}e^{\lambda_2\int_0^\infty K(s)ds},$$

where $x_0^e = (\hat{x}_0, e_0)$.
Hence,

$$\|\hat{x}(t)\| \leq \sqrt{\frac{1}{\alpha}}\left[\max(\sqrt{\alpha\|P_1\|}, \sqrt{\|P_2\|})(\|\hat{x}_0\| + \|e_0\|)e^{-\frac{\lambda_1}{2}(t-t_0)}e^{\frac{\lambda_2}{2}\int_0^\infty K(s)ds}\right].$$

Consequently, the cascade system (14) is globally uniformly exponentially stable. This ends the proof. □

4 Example

We Consider the controlled metal bar equation

$$\begin{cases} \dfrac{\partial x(\zeta,t)}{\partial t} = \dfrac{\partial^2 x(\zeta,t)}{\partial^2\zeta} + b(\zeta)u(t) + \dfrac{\sin(x(\zeta,t))}{1+t^2}, \\ \dfrac{\partial x}{\partial\zeta}(0,t) = 0 = \dfrac{\partial x}{\partial\zeta}(1,t), \quad x(\zeta,0) = x_0(\zeta), \quad t \geq 0, \\ y(t) = \displaystyle\int_0^1 c(\zeta)x(\zeta,t)d\zeta, \end{cases} \tag{15}$$

where $x(\zeta, t)$ represents the temperature at position ζ at time $t \geq 0$ and x_0 represents the initial temperature profile, $u(t)$ the addition of heat along the bar and b, c represents the shaping functions around the control ζ_0 and the sensing point ζ_1, respectively

$$b(\zeta) = \frac{1}{2\delta}\mathbf{1}_{[\zeta_0-\delta,\zeta_0+\delta]}$$

and

$$c(\zeta) = \frac{1}{2\kappa}\mathbf{1}_{[\zeta_1-\kappa,\zeta_1+\kappa]},$$

with $[\zeta_0 - \delta, \zeta_0 + \delta] \cap [\zeta_1 - \kappa, \zeta_1 + \kappa] = \emptyset$.

Notice that b and c in this example are both elements in $L^2(0, 1)$ for a fixed small, non-negative constants δ and κ.

The partial differential equation is equivalent to system (1) where $H = L^2(0, 1)$, $U = \mathbb{C}$, $Y = \mathbb{C}$, $A = \dfrac{\partial^2}{\partial^2\zeta}$, with

$$D(A) = \{h \in L^2(0, 1), h, \tfrac{\partial h}{\partial \zeta} \text{ are absolutely continuous}, \tfrac{\partial^2 h}{\partial \zeta^2} \in L^2(0, 1) \text{ and } \tfrac{dh}{d\zeta}(0) = 0 = \tfrac{dh}{d\zeta}(1)\},$$

The input operator

$$Bu = b(\zeta)u,$$

$B \in L(\mathbb{C}, H)$, and has norm $\frac{1}{\sqrt{2\delta}}$.

Besides, the measured output operator

$$Cx = \int_0^1 c(\zeta)x(\zeta, t)d\zeta,$$

where $C \in L(H, \mathbb{C})$, and has norm $\frac{1}{\sqrt{2\kappa}}$, and

$$F(t, x) = \frac{\sin(x(\zeta, t))}{1 + t^2}.$$

A has the eigenvalues $0, -n^2\pi^2, \ n \geq 1$ and the corresponding orthogonal eigenvectors are

$$v_n = \begin{cases} 1 \ \text{if} \ n = 0, \\ \sqrt{2}\cos(n\pi\zeta) \ \text{if} \ n \geq 1. \end{cases}$$

It follows that, A is the infinitesimal generator of the C_0−semigroup (see Curtain and Zwart [9] for details).

We can take as a stabilizing feedback $u(t) = Dx$ with

$$Dx = -3\langle x, v_0\rangle = -3\langle x, 1\rangle,$$

where $\langle , \rangle$ is the inner product on $L^2(0, 1)$.

It is easy to verify that $A + BD$ has the eigenvalues $-3, -(n\pi)^2, n \geq 1$. Then, the pair $\{A, B\}$ is exponentially stabilizable.

In addition, the stabilizing output injection is given by

$$Ly = -3yv_0 = -3y.1.$$

The system $A + LC$ has the eigenvalues $-3, -(n\pi)^2, n \geq 1$. Then, the pair $\{A, C\}$ is exponentially detectable.

On the other hand, the assumption $(\mathcal{H}_1)$ is satisfied with $K(t) = \frac{1}{1+t^2}$ which is non-negative continuous and integrable on $\mathbb{R}_+$.

Thus, all assumptions of Theorem 3 are satisfied. We conclude that a stabilizing compensator is given by

$$\begin{cases} \dfrac{\partial \hat{x}(\zeta, t)}{\partial t} = \dfrac{\partial^2 \hat{x}(\zeta, t)}{\partial \zeta^2} - \dfrac{3}{2\kappa} \displaystyle\int_{\zeta_1-\kappa}^{\zeta_1+\kappa} \hat{x}(\zeta, t) d\zeta + \dfrac{1}{2\delta} \mathbf{1}_{[\zeta_0-\delta, \zeta_0+\delta]}(\zeta) u(t) + \dfrac{\sin(\hat{x}(\zeta, t))}{1+t^2} + 3y(t), \\ \dfrac{\partial \hat{x}}{\partial \zeta}(0, t) = 0 = \dfrac{\partial \hat{x}}{\partial \zeta}(1, t), \quad \hat{x}(\zeta, 0) = \hat{x}_0(\zeta), \quad t \geq 0, \\ u(t) = -3 \displaystyle\int_0^1 \hat{x}(\zeta, t) d\zeta. \end{cases} \tag{16}$$

5 Conclusion

We have presented the problems of state observation and state trajectory control via output feedback for a class of non-linear system. It is shown that the system can be stabilizable by means of an estimated state feedback given by a designated Luenberger observer. Furthermore, a compensator design of a class of nonlinear uncertain evolution equations on a Hilbert space has been considered. An example has been introduced to show the applicability of our theoretical results.

References

1. Benabdallah, A.: A separation principle for the stabilization of a class of time-delay nonlinear systems. Kybernetika **51** 99–111 (2015)
2. Benabdallah, A., Ellouze, I., Hammami, M.A.: Practical exponential stability of perturbed triangular systems and a separation principle. Asian J. Control **13** 445–448 (2011)

3. Bernstein, D.S., Hyland, D.C: The optimal projection equations for finite-dimensional fixed-order dynamic compensation of infinite-dimensional systems. Siam J. Control Optim. **24** 122–151 (1986)
4. Curtain, R.F.: Compensators for infinite dimensional linear systems. J. Franklin Inst. **315** 331–346 (1983)
5. Curtain, R.F.: Finite-dimensional compensators for parabolic distributed systems with unbounded control and observation. SIAM J. Control Optim. **22** 225–276 (1984)
6. Curtain, R.F.: A comparaison of finite-dimensional controller designs for distributed parameter systems. Control Theory Technol. **9** 609–628 (1993)
7. Curtain, R.F.: Stabilization of boundary control distributed systems via integral dynamic output feedback of a finite-dimensional compensator. Anal Optim Syst **44** 761–776 (1982)
8. Curtain. R.F., Salamon, D.: Finite-dimensional compensators for infinite-dimensional systems with unbounded input operators. SIAM J. Control Optim. **24** 797–816 (1986)
9. Curtain, R.F., Zwart, H.J.: An Introduction to Infinite Dimensional Linear Systems Theory. Springer, New York (1995)
10. Damak, H., Hammami, M.A.: Stabilization and practical asymptotic stability of abstract differential equations. Numer. Funct. Anal. Optim. **37** 1235–1247 (2016)
11. Damak, H., Ellouze, I., Hammami, M.A.: A separation principle of a class of time-varying nonlinear systems. Nonlinear Dyn. Syst. Theory **13** 133–143 (2013)
12. Damak, H., Hammami, M.A.: Stabilization of a class of time-varying systems in Hilbert spaces. Springer Proc. Math. Stat. **131** 263–269 (2015)
13. Damak, H., Hammami, M.A.: Asymptotic stability of a perturbed abstract differential equations in Banach spaces. Oper. Matrices **14** 129–138 (2020)
14. Damak, H.: Compensator design via the separation principle for a class of semilinear evolution equations. Accepted in Ukr. Math. J.
15. Damak, H.: The practical feedback stabilization for evolution equations in banach spaces. Acta Mech. et Autom. **15** 58–65 (2021)
16. Ellouze, I., Hammami, M.A.: A separation principle of time-varying dynamical systems: a practical stability approach. Math. Model. Anal. **12** 297–308 (2007)
17. Gressang, R., Lamont, G.: Observers for systems characterized by semigroups. IEEE Trans. Autom. Control **AC-20** 523–528 (1975)
18. Jmal, A., Naifar, O., Ben Makhlouf, A., Derbel, N., Hammami, M.A.: On observers design for nonlinear caputo fractional-order systems. Asian J. Control **20** 1–8 (2018)
19. Kitamura, S., Sakairi, H., Mishimura, M.: Observers for disturbuted parameter systems. Electri Eng. Jpn. **92** 142–149 (1972)
20. Loria, A., Panteley, E.: A separation principle for a class of Euler-Langrange systems. In: Nijmeijer, H., Fossen, T. (eds.) New Directions in Nonlinear Observer Design, Lecture Notes in Control and Information Sciences, vol 244. Springer, London (1999)
21. Orlov, Y., Lou, Y., Christofides, P.D.: Robust stabilization of infinite-dimensional systems using sliding-mode output feedback control. Int. J. Control **77** 1115–1136 (2004)
22. Naifar, O., Makhlouf, A.B., Hammami, M.A.: Global practical mittag leffler stalilization output feedback for a class of nonlinear fractional-order systems. Asian J. Control **20** 1–9 (2018)
23. Pazy, A.: Semigroups of Linear Operators and Applications to Partial Differential Equations. Springer, New York, NY, USA (1983)
24. Qiang, Y., Wang, J.W., Sun, C.Y.: Observer-based output feedback compensator design for linear parabolic PDEs with local piecewise control and pointwise observation in space. IET Control Theory Appl. **12** 1812–1821 (2018)
25. Sakawa, Y., Matsushita, T.: Feedback stabilization for a class of distributed systems and construction of a state estimator. IEEE Trans. Autom. Control **AC-20** 748–753 (1975)
26. Teschl, G.: Ordinary Differential Equations and Dynamical Systems, Graduate studies in mathematics. American Mathematical Society (2012)

Observer Desing for Non Linear Takagi-Sugeno Fuzzy Systems. Application to Fault Tolerant Control

Atef Khedher, Ilyes Elleuch, and Kamal Ben Othman

Abstract The problem of fuzzy fault tolerant control design for systems described by Takagi-Sugeno models is studied in this chapter. The fault tolerant control design require the state and fault estimation. In order to make this estimation, a proportional integral observer is conceived. The proposed method shows that it is possible to conceive simultaneously the proportional integral observer and the fuzzy fault tolerant control. The cases of system affected by actuator and/or sensor faults are considered. In order to conceive the fault tolerant control strategy for the case of sensor faults, a mathematical transformation is used allowing conceiving an augmented system, in which the initial sensor fault appears as an actuator fault. The fault tolerant control and the proportional integral observer are both conceived considering the augmented state. The noise effect on the state and fault estimation is also minimized in this study, which provides some robustness properties to the proposed control and observer. The fault tolerant control and proportional integral observer design is formulated in term of linear matrix inequalities (LMI).

Keywords Fuzzy fault tolerant control · Proportional integral observer · Actuator faults · Faults estimation · Takagi-Sugeno fuzzy systems · Sensor faults · State estimation · Linear matrix inequalities

1 Introduction

Fuzzy systems have been studied for many years since their first definition by Zadeh in 1965 [47]. They are considered in the context of control [5], identification [40], state estimation [3] and diagnosis [37]. Many complex process can be modeled using

A. Khedher (✉)
LARA Automatique, ENIT, Tunis, Tunisia

I. Elleuch
University of Sousse, Sousse, Tunisia

K. B. Othman
LASEE, ENIM, Monastir, Tunisia

O. Naifar and A. Ben Makhlouf (eds.), *Advances in Observer Design and Observation for Nonlinear Systems*, Studies in Systems, Decision and Control 410,
https://doi.org/10.1007/978-3-030-92731-8_7

fuzzy sets concept [17]. Among the different kinds of fuzzy systems, Takagi-Sugeno models have been considered as a good framework of non linear systems modeling [4, 13, 18, 27, 28, 37, 48]. They are used as a tool of modeling of non linear systems since their simplicity of manipulation and their particular expression [42, 43].

In many cases, processes can be affected by disturbances due to the noises. Moreover, It is possible that system sensors and/or actuators can be corrupted by faults or failures. In certain cases, it is impossible to repair the fault affecting the system for many reasons, such that the constraint of time, or the conditions of production, etc... In that cases, it is possible to use control strategy able to minimize their effects. This control strategy is named fault tolerant control (FTC).

A control loop is considered as a fault tolerant if there exist adaptation strategies of the initial control law included in the closed-loop that introduce redundancy in actuators [45]. The idea of fault Tolerant Control can be considered as a new idea in the literature [8]. It allows to have a control low that fulfills its objectives when fault appears [28].

There exist two main kinds of fault tolerant control strategies: the active and the passive ones. The passive techniques are control laws that take into account all the possible faults since the system design. This kind of control techniques is described in [8, 11, 32, 33, 39]. The active fault tolerant control strategies allows adapting the control law using the information provided by the FDI block [7, 8, 28, 45]. So the design of an active fault tolerant control require the existence of a FDI block in the process. Active fault tolerant control has been the subject of many researches, in the last years [19, 24, 26, 30, 44].

In the context of Takagi-Sugeno systems, Many works have interested in the design of different fault tolerant control lows [1, 15, 24, 44]. In [44] a fault tolerant control low is conceived in the case of system affected by sensor fault. A predictive fault tolerant Control is conceived for LPV systems based on model reference in [1]. In [24], authors used an unknown observer to conceive the fault tolerant control. State and fault estimation was applied to the fault tolerant control in [16, 34, 36, 41].

In this chapter, an active FTC strategy is proposed. The proposed FTC is designed on the base of Takagi-Sugeno fuzzy concept and it is implemented as a state feedback controller. This FTC depends on the state estimation error and the fault estimate. In this conditions, the state and fault estimation becomes a necessity, so a proportional integral observer is conceived in order to estimate simultaneously the system state and the fault.

The main goal of this chapter is to propose simultaneously a fuzzy fault tolerant control and a proportional integral observer(PIO). The proposed control is conceived in the case of systems affected by actuator fault firstly. In second case system affected by sensor fault are considered. Finally, the case where systems are affected by sensor and actuator faults is studied. In the cas where system is affected by sensor fault, a mathematical transformation [13, 18, 27] based on the definition of an augmented stated is used. Considering this augmented state, the sensor fault appears as an actuator fault. The FTC and the PIO are conceived on the base of the augmented system.

The chapter is organized as follows. Section 2 recalls the principle of Takagi-Sugeno multiple models. Section 3 presents the principle of the proportional integral observer design for Takagi-Sugeno models allowing state and fault estimation. Section 4 summarizes the principle of the simultaneous proportional integral observer and fault tolerant control design based on the separation principle. Section 5 describes the design of the fuzzy fault tolerant control and the proportional integral observer in the case of system affected by actuator faults. The design of the FTC and the PIO is adapted to system affected by sensor faults in Sect. 6. Section 7 studied the case of system affected, simultaneously, by sensor and actuator faults. An example of simulation showing the robustness of the proposed method is is given in Sect. 8.

2 Takagi-Sugeno Fuzzy Systems

Takagi-Sugeno fuzzy systems are an appropriate tool which permits to model large class of complex and non linear systems with a mathematical model which can be used for analysis [10, 12], control [6, 31] and observer design [9, 18, 27, 35]. This approach is based on a decomposition of the system operating space into a finite number of operating zones. Hence, a simple linear model describes the system dynamic behavior inside each operating zone. The contribution of each sub-model in the global model is quantified using a non linear weighting function which can have various structures. The sub-models are associated in the state equation using a common state vector. This model has been proposed, in a fuzzy modeling framework, by Takagi and Sugeno [42] and in a multiple model modeling framework by Johansen and Foss [22].

Takagi-Sugeno fuzzy systems is based on the assumption that each nonlinear dynamic system can be simply, described the fuzzy fusion of many linear models. where each linear model represents the local system behavior around an operating point. A Takagi-Sugeno model is described by fuzzy IF-THEN rules which represent local linear Inputs/Outputs relations of the non-linear system. It has a rule base of M rules, each having p antecedents, where the ith rule is expressed as follows:

$$\begin{aligned} R^i &: \text{IF } \xi_1 \text{ is } F_1^i \text{ and ... and } \xi_p \text{ is } F_p^i \\ &\text{THEN}: \begin{cases} \dot{x}(t) = A_i x(t) + B_i u(t) \\ y = C_i x(t) \end{cases} \end{aligned} \tag{1}$$

where: $i \in \{1, ..., M\}$, $F_j^i (j \in \{1, ..., p\})$ are fuzzy sets and $\xi = [\xi_1, \xi_2, ..., \xi_p]$ is a known vector of premise variables [29] which may depend on the state, the input or the output. Variables ξ is called decision variable.

The global Takagi-Sugeno fuzzy model is given by the aggregation of the sub-models using the weighting functions as follows:

$$\begin{cases} \dot{x}(t) = \sum_{i=1}^{\ell} \mu_i(\xi(t)) \left(A_i x(t) + B_i u(t)\right) \\ y(t) = \sum_{i=1}^{\ell} \mu_i(\xi(t)) C_i x(t) \end{cases} \tag{2}$$

where $x(t) \in \mathbb{R}^n$ is the state vector, $u(t) \in \mathbb{R}^r$ is the control vector, $y(t) \in \mathbb{R}^m$ is the vector of measures and A_i, B_i and C_i are known constant matrices with appropriate dimensions.

The weighting functions $\mu_i(\xi(t))$ assure a progressive passage between the local models and verifies the following convex proprieties: $\sum_{i=1}^{\ell} \mu_i(\xi(t)) = 1, \ \forall t$ and $0 \leq \mu_i(\xi(t)) \leq 1, \ \forall i = 1...\ell, \ \forall t$

If, in the equation of the output, it is supposed that $C_1 = C_2 = ... = C_\ell = C$, the output of the multiple model (2) is reduced to : $y(t) = Cx(t)$ and the multiple model state equation becomes:

$$\begin{cases} \dot{x}(t) = \sum_{i=1}^{\ell} \mu_i(\xi(t))(A_i x(t) + B_i u(t)) \\ y(t) = Cx(t) \end{cases} \tag{3}$$

3 Observers Design for Non Linear Takagi-Sugeno Fuzzy Systems

3.1 Introduction

Takagi-Sugeno models are considered in many works [4, 23, 40, 48], in the context of state estimation in spite of parameter uncertainties of unknown inputs. Let us present briefly some works in this context. In [4], a Singulary perturbed Takagi-Sugeno models is presented. For this system, activation function depend on unmeasurable variables which can be system state for example. In [40], authors propose an apporach based on an online identification using amodel of Takagi-Sugeno-Kang applied to crane systems. The problem of sensor networks for nonfragile distributed filters and its application to Takag-Sugeno models was studied in [48]. In [13], authors proposed a new form of observer able to estimate fault and system state for uncertain systems modeld by Taagi-Sugeno models based on the sliding mode principle. State and fault estimation for non linear systems described by Takag-Sugeno fuzzy models are consided in [27, 37].

State and fault estimation can be made generally using proportional integral observers. This observer contains two terms called proportional and integral terms. The proportional term is used to estimate system state and the integral term allows estimating the faults which are considered as unknown inputs [13, 18, 27, 37]. In practice, proportional integral observer is characterized by two gains (proportional and integral) [13, 18, 27, 37]. Works presented in [13, 18, 27, 37] present the design of this kind of observer with many academic and real applications for state and fault estimation.

State and fault estimation is studied also in nuerous works in many different contexts. Multi agent systems are considered for state and fault estimation in [20, 25, 46]. In [21], authors propose to make the aymptotic fault and state estimation and fault tolerant control design for the case of non linear systems. In [38], authors assume that the activation function verifies Lipschitz condition. Several works are interested in Discrete Takagi-Sugeno models [2, 14, 49].

3.2 Proportional Integral Observer Design

The main goal of this section is to recall the principle of the proportional integral observer design for the state and fault estimation in the case of Takagi Sugeno models.

Consider the Takagi-Sugeno fuzzy system affected by an actuator fault and a measurement noise given by the following equation:

$$\begin{cases} \dot{x}(t) = \sum_{i=1}^{\ell} \mu_i(\xi(t))(A_i x(t) + B_i u(t) + E_i f_a(t)) \\ y(t) = Cx(t) + Dw(t) \end{cases} \tag{4}$$

where $x(t) \in \mathbb{R}^n$ is the system state, $y(t) \in \mathbb{R}^m$ is the measured output, $u(t) \in \mathbb{R}^r$ is the input, $f_a(t)$ is the actuator fault which is assumed to be bounded. and $w(t)$ is the measurement noise. A_i, B_i and C are known constant matrices with appropriate dimensions. E_i and D are respectively the fault and noise distribution matrices which are assumed to be known. The scalar M represents the number of the local models. $\mu_i(\xi(t))$ are the activation functions.

The considered proportional integral observer is presented as follows:

$$\begin{cases} \dot{\hat{x}}(t) = \sum_{i=1}^{\ell} \mu_i(\xi(t))(A_i \hat{x}(t) + B_i u(t) + E_i \hat{f}_a(t) + K_i \tilde{y}(t)) \\ \dot{\hat{f}}_a(t) = \sum_{i=1}^{\ell} \mu_i(\xi(t)) L_i \tilde{y}(t) \\ \hat{y}(t) = C\hat{x}(t) \end{cases} \tag{5}$$

where $\hat{x}(t) \in \mathbb{R}^n$ is the estimated system state, $\hat{f}_a(t)$ represents the estimated fault, $\hat{y}(t) \in \mathbb{R}^m$ is the estimated output, $\tilde{y}(t) = y(t) - \hat{y}(t)$, K_i are the local proportional observer gains and L_i are the local integral gains to be computed.

The state estimation error $\tilde{x}(t)$ and the fault estimation error $\tilde{f}_a(t)$ are defined as following:

$$\tilde{x}(t) = x(t) - \hat{x}(t) \tag{6}$$

$$\tilde{f}_a(t) = f_a(t) - \hat{f}_a(t) \tag{7}$$

The dynamics of the state estimation error is given by the computation of $\dot{\tilde{x}}(t)$:

$$\begin{aligned} \dot{\tilde{x}}(t) &= \dot{x}(t) - \dot{\hat{x}}(t) \\ &= \sum_{i=1}^{\ell} \mu_i(\xi(t))(A_i - K_i C)\tilde{x}(t) + E_i \tilde{f}_a(t) + K_i Dw(t)) \end{aligned} \tag{8}$$

The dynamics of the fault estimation error is given by the expression of $\dot{\tilde{f}}_a(t)$:

$$\begin{aligned}\dot{\tilde{f}}_a(t) &= \dot{f}_a(t) - \dot{\hat{f}}_a(t) \\ &= \dot{f}_a(t) - \sum_{i=1}^{\ell} \mu_i(\xi(t))(L_i C\tilde{x}(t) - L_i Dw(t))\end{aligned} \tag{9}$$

The following matrices can be introduced:

$$\varphi_a = \begin{bmatrix} \tilde{x}(t) \\ \tilde{f}_a(t) \end{bmatrix} \text{ and } \varepsilon_a = \begin{bmatrix} w(t) \\ \dot{f}_a(t) \end{bmatrix} \tag{10}$$

Using theses matrices, the Eqs. (8) and (9) can be reformulated as follows:

$$\dot{\varphi}_a = A_{ma}\varphi_a + B_{ma}\varepsilon_a \tag{11}$$

with:

$$A_{ma} = \sum_{i=1}^{\ell} \mu_i(\xi(t))A_{ai} \text{ and } B_{ma} = \sum_{i=1}^{\ell} \mu_i(\xi(t))B_{ai} \tag{12}$$

where:

$$A_{ai} = \begin{bmatrix} A_i - K_i C & E_i \\ -L_i C & 0 \end{bmatrix} \text{ and } B_{ai} = \begin{bmatrix} -K_i D & 0 \\ -L_i D & I \end{bmatrix} \tag{13}$$

The matrix I is the identity matrix with appropriate dimensions.

To analyze the convergence of the generalized estimation error $\varphi_a(t)$, the Lyapunov function $V_a(t) = \varphi_a(t)^T P \varphi_a(t)$ is considered, where P is a symmetric definite positive matrix.

The problem of state and fault estimation is based on the computation of the gains K_i and L_i of the observer to ensure an asymptotic convergence of $\varphi_a(t)$ toward zero if $\varepsilon_a(t) = 0$ and bounded error in the case where $\varepsilon_a(t) \neq 0$, i.e.:

$$\begin{aligned} &\lim_{t\to\infty} \varphi_a(t) = 0 && \text{for } \varepsilon_a(t) = 0 \\ &\|\varphi_a(t)\|_{Q_\varphi} \le \lambda \|\varepsilon_a(t)\|_{Q_\varepsilon} && \text{for } \varepsilon_a(t) \neq 0 \text{ and } e(0) = 0 \end{aligned} \tag{14}$$

where $\lambda > 0$ is the attenuation level.

To satisfy the constraints (14), it is sufficient to find a Lyapunov function $V_a(t)$ verifying:

$$\dot{V}_a(t) + \varphi_a^T Q_\varphi \varphi_a - \lambda^2 \varepsilon_a^T Q_\varepsilon \varepsilon_a < 0 \tag{15}$$

Q_φ and Q_ε are two positive definite matrices.

In order to simplify the notations, the time index (t) will be omitted henceforth. Inequality (15) can also be written as follows:

$$\psi_a^T \Omega_a \psi_a < 0 \tag{16}$$

with:

$$\psi_a = \begin{bmatrix} \varphi_a \\ \varepsilon_a \end{bmatrix} \text{ and } \Omega_a = \begin{bmatrix} A_{ma}^T P + P A_{ma} + Q_\varphi & P B_{ma} \\ B_{ma}^T P & -\lambda^2 Q_\varepsilon \end{bmatrix} \tag{17}$$

Inequality (16) has a quadratic form, it holds iff $\Omega_a < 0$.

The matrices A_{ma} and B_{ma} can be written as :

$$A_{ma} = \tilde{A}_{ma} - \tilde{K}_{ma}\tilde{C} \text{ and } B_{ma} = \tilde{I} - \tilde{K}_{ma}\tilde{D} \tag{18}$$

where:

$$\tilde{A}_{ma} = \sum_{i=1}^{\ell} \mu_i(\xi(t))\tilde{A}_{mai}, \quad \tilde{K}_{ma} = \sum_{i=1}^{\ell} \mu_i(\xi(t))\tilde{K}_{mai} \tag{19}$$

with:

$$\tilde{K}_{mai} = \begin{bmatrix} K_i \\ L_i \end{bmatrix}, \quad \tilde{A}_{mai} = \begin{bmatrix} A_i & E_i \\ 0 & 0 \end{bmatrix}, \quad \tilde{I} = \begin{bmatrix} 0 & 0 \\ 0 & I \end{bmatrix}$$
$$\tilde{C} = \begin{bmatrix} C & 0 \end{bmatrix}, \text{ and } \tilde{D} = \begin{bmatrix} D & 0 \end{bmatrix},$$

With the changes of variables $G_{ma} = P\tilde{K}_{ma}$ and $\bar{\lambda} = \lambda^2$, the matrix Ω_a can be put as following:

$$\Omega_a = \begin{bmatrix} \vartheta_a & -G_{ma}\tilde{D} + P\tilde{I} \\ \tilde{I}^T P - \tilde{D}^T G_{ma}^T & -\bar{\lambda} Q_\varepsilon \end{bmatrix} \tag{20}$$

where $\vartheta_a = P\tilde{A}_{ma} + \tilde{A}_{ma}^T P - G_{ma}\tilde{C} - \tilde{C}^T G_{ma}^T + Q_\varphi$.

As $\Omega_a = \sum_{i=1}^{\ell} \mu_i(\xi(t))\Omega_{ai}$, the negativity of Ω is assured if, for $i = 1...\ell$:

$$\Omega_{ai} < 0 \tag{21}$$

with:

$$\Omega_{ai} = \begin{bmatrix} \vartheta_{ai} & -G_{ai}\tilde{D} + P\tilde{I} \\ \tilde{I}^T P - \tilde{D}^T G_{ai}^T & -\bar{\lambda} Q_\varepsilon \end{bmatrix} \tag{22}$$

where: $\vartheta_{ai} = P\tilde{A}_{mai} + \tilde{A}_{mai}^T P - G_{ai}\tilde{C} - \tilde{C}^T G_{ai}^T + Q_\varphi$ and $G_{ai} = P\tilde{K}_{mai}$.

The resolution of the LMI's (21) leads to the determination of the matrices P and G_{ai} and the scalar $\bar{\lambda}$. The gain matrices are then deduced by the equation $\tilde{K}_{mai} = P^{-1}G_{ai}$.

The observer design is summarized by the following theorem:

Theorem 1 *The system (11) describing the time evolution of the state estimation error $\tilde{x}$ and the fault estimation error $\tilde{f}_a$ is stable and the $\mathcal{L}_2$–gain of the transfer from $\varepsilon_a(t)$ to $\varphi_a(t)$ is bounded, if there exists a symmetric, positive definite matrix P, gain matrices G_{ai}, $i \in \{1...\ell\}$ and a positive scalar $\bar{\lambda}$ such that the following LMI are verified:*

$$\begin{bmatrix} \vartheta_{ai} & -G_{ai}\tilde{D} + P\tilde{I} \\ \tilde{I}^T P - \tilde{D}^T G_{ai}^T & -\bar{\lambda} Q_{\varepsilon} \end{bmatrix} < 0 \quad i \in \{1...\ell\} \tag{23}$$

where: $\vartheta_{ai} = P\tilde{A}_{mai} + \tilde{A}_{mai}^T P - G_{ai}\tilde{C} - \tilde{C}^T G_{ai}^T + Q_{\varphi}$. The observer gains (proportional and integral gains) are computed using $\tilde{K}_{mai} = P^{-1}G_{ai}$ and the attenuation level is given by $\lambda = \sqrt{\bar{\lambda}}$. ■

4 Observer and Fault Tolerant Control Design Using Separation Principle

The objective of this section is to recall the method of simultaneous proportional integral observer and the fault tolerant control using the separation principle in the context of non linear systems described by Takagi-Sugeno models.

A non linear system described by Takagi-Sugeno model can be expressed as follow :

$$\dot{x}(t) = \sum_{i=1}^{\ell} \mu_i(u(t))A_i x(t) + Bu(t) \tag{24a}$$

$$y(t) = Cx(t) \tag{24b}$$

where $x(t) \in R^n$ is the state vector, $u(t) \in R^r$ is the input vector, $y(t) \in R^m$ the output vector and A_i, B_i and C_i are known constant matrices with appropriate dimensions. The scalar M represents the number of local models.

Consider the following nonlinear Takagi-Sugeno model affected by actuator faults and measurement noise:

$$\dot{x}_f(t) = \sum_{i=1}^{\ell} \mu_i(u(t))A_i x_f(t) + Bu_f(t) + Ef_a(t) \tag{25a}$$

$$y_f(t) = Cx_f(t) + Dw(t) \tag{25b}$$

where $x_f(t) \in R^n$ is the state vector, $u_f(t) \in R^r$ is the input vector, $y_f(t) \in R^m$ the output vector. $f_a(t)$ represents the fault which is assumed to be bounded and $w(t)$ is the measurement noise. E and D are respectively the fault and the noise distribution matrices which are assumed to be known. The scalar M represents the number of local models.

The structure of the proportional integral observer is chosen as follows:

$$\dot{\hat{x}}_f(t) = \sum_{i=1}^{\ell} \mu_i(u(t))(A_i \hat{x}_f(t) + K_i(y_f(t) - \hat{y}_f(t))) + Bu_f(t) + E\hat{f}_a(t) \quad (26a)$$

$$\dot{\hat{f}}_a(t) = \sum_{i=1}^{\ell} \mu_i(u(t))(L_i(y_f(t) - \hat{y}_f(t))) \quad (26b)$$

$$\hat{y}_f(t) = C\hat{x}_f(t) \quad (26c)$$

where $\hat{x}_f(t)$ is the estimated system state, $\hat{f}_a(t)$ represents the estimated fault, $\hat{y}_f(t)$ is the estimated output, K_i are the local model proportional observer gains and L_i are the local model integral gains to be computed.

The fault tolerant control $u_f(t)$ is described by the following expression :

$$u_f(t) = -S\hat{f}_a(t) + M(x(t) - \hat{x}_f(t)) + u(t) \quad (27)$$

where S and M are two constant matrices with appropriate dimensions. The objective is to find the matrices S and M which permit to the state x_f to converge to x.

Let us define $\tilde{x}(t)$ the error between the states $x(t)$ and $x_f(t)$, $\tilde{x}_f(t)$ the estimation error of the state x_f and $\tilde{f}_a(t)$ the fault estimation error :

$$\tilde{x}(t) = x(t) - x_f(t) \quad (28)$$

$$\tilde{x}_f(t) =_f (t) - \hat{x}_f(t) \quad (29)$$

$$\tilde{f}_a(t) = f_a(t) - \hat{f}_a(t) \quad (30)$$

Using the control strategy described by the Eq. (27), the dynamics of the errors defined in (28), (29) and (30) can be written as follow :

$$\dot{\tilde{x}}(t) = (\sum_{i=1}^{\ell} \mu_i(u(t))A_i - BM)\tilde{x}(t) + BS\hat{f}_a(t)$$
$$-BM\tilde{x}_f(t) - Ef_a(t) \quad (31)$$

Choosing S verifying $E = BS$, the dynamics of $\tilde{x}(t)$ becomes :

$$\dot{\tilde{x}}(t) = (\sum_{i=1}^{\ell} \mu_i(u(t))A_i - BM)\tilde{x}(t) - E\tilde{f}_a(t)) - BM\tilde{x}_f(t) \quad (32)$$

The dynamic of $\tilde{x}_f(t)$ can be written :

$$\dot{\tilde{x}}_f(t) = (\sum_{i=1}^{\ell} \mu_i(u(t))(A_i - K_i C)\tilde{x}_f(t) - K_i Dw(t)) + E\tilde{f}_a(t) \tag{33}$$

The dynamic of the fault error estimation can be written :

$$\dot{\tilde{f}}_a(t) = \dot{f}_a(t) - \sum_{i=1}^{\ell} \mu_i(u(t))(L_i C\tilde{x}_f + L_i Dw(t)) \tag{34}$$

In order to simplify the notations, the time index (t) will be omitted henceforth.

The Eqs. (32), (33) and (34) can be rewritten :

$$\dot{\varphi} = A_m \varphi + B_m \psi \tag{35}$$

with :

$$\varphi = \begin{bmatrix} \tilde{x}^T \tilde{x}_f^T \tilde{f}_a^T \end{bmatrix}^T \text{ and } \psi = \begin{bmatrix} w^T \dot{f}_a^T \end{bmatrix}^T \tag{36}$$

and :

$$A_m = \sum_{i=1}^{\ell} \mu_i(u(t))A_{mi} \text{ and } B_m = \sum_{i=1}^{\ell} \mu_i(u(t))B_{mi} \tag{37}$$

where :

$$A_{mi} = \begin{bmatrix} A_i - BM & 0 & -E \\ 0 & A_i - K_i C & E \\ 0 & -L_i C & 0 \end{bmatrix} \tag{38}$$

and

$$B_{mi} = \begin{bmatrix} 0 & 0 \\ -K_i D & 0 \\ -L_i D & I \end{bmatrix} \tag{39}$$

Considering the Lyapunov $V(t) = \varphi^T(t)P\varphi(t)$ where P denotes a positive definite matrix, the errors converge to zero if $\dot{V} < 0$. $\dot{V} < 0$ if $A_{mi}^T P + P A_{mi} < 0, \quad \forall i \in \{1, ..., \ell\}$.

The matrices A_{mi} and B_{mi} can be rewritten :

$$A_{mi} = \begin{bmatrix} A_i - BM & E_1 \\ 0 & \tilde{A}_i - \tilde{K}_i \tilde{C} \end{bmatrix} \tag{40}$$

and

$$B_{mi} = \begin{bmatrix} 0 \\ \tilde{I} - \tilde{K}_i \tilde{D} \end{bmatrix} \tag{41}$$

with :

$$\tilde{A}_i = \begin{bmatrix} A_i & E \\ 0 & 0 \end{bmatrix}, \quad \tilde{K}_i = \begin{bmatrix} K_i \\ L_i \end{bmatrix}, \quad \tilde{I} = \begin{bmatrix} 0 & 0 \\ 0 & I \end{bmatrix}$$
$$\tilde{C} = \begin{bmatrix} C & 0 \end{bmatrix}, \quad \tilde{D} = \begin{bmatrix} D & 0 \end{bmatrix} \text{ and } E_1 = \begin{bmatrix} -BM & -E \end{bmatrix}$$

Assuming that P has the block diagonal form $P = diag(P_1, P_2)$, and using the separation principle, φ converges to zero iff there exist matrices $P_1 > 0$ and $P_2 > 0$ such that following inequalities are satisfied:

$$(A_i - BM)^T P_1 + P_1(A_i - BM) < 0 \tag{42}$$
$$(\tilde{A}_i - \tilde{K}_i\tilde{C})^T P_2 + P_2(\tilde{A}_i - \tilde{K}_i\tilde{C}) < 0 \tag{43}$$

By multiplying (42) from left and right by P_1^{-1} one obtain :

$$P_1^{-1}(A_i - BM)^T + (A_i - BM)P_1^{-1} < 0 \tag{44}$$

Substituing $W = P_1^{-1}$, the Eq. (44) becomes:

$$W(A_i - BM)^T + (A_i - BM)W < 0 \tag{45}$$

φ converge to zero if there exist two definites and posistives mtrices W and P_2 satisfaying (43) and (45)

The inegalities (43) and (45) are not linears, substituing $X = MW$, and $Y_i = P_2\tilde{K}_i$, their become:

$$WA^T + AW - X^T B^T - BX < 0 \tag{46}$$
$$\tilde{A}^T P_2 + P_2\tilde{A} - Y\tilde{C} - \tilde{C}^T Y^T < 0 \tag{47}$$

The resolution of the linear matrices inegalities (LMI) (46) and (47) permits to find the matrices W, P_2, X and Y_i.

The matrices M and $\tilde{K}$ are computed using the following equations:

$$M = XW^{-1} \tag{48}$$

$$\tilde{K}_i = P_2^{-1} Y_i \tag{49}$$

5 Fuzzy Fault Tolerant Control Design for System with Actuator Fault

A non linear system described by Takagi-Sugeno fuzzy structure can be expressed as follows:

$$\begin{cases} \dot{x}(t) = \sum\limits_{i=1}^{\ell} \mu_i(u(t))A_i x(t) + Bu(t) \\ y(t) = Cx(t) \end{cases} \tag{50}$$

where $x(t) \in \mathbb{R}^n$ is the state vector, $u(t) \in \mathbb{R}^q$ is the system input vector, $y(t) \in \mathbb{R}^m$ is the measured output vector and A_i, B and C are the system matrices wich are known and constant with appropriate dimensions. The scalar r is the number of local models.

A nonlinear Takagi-Sugeno model affected by actuator faults and measurement noise is given by the following state equation:

$$\begin{cases} \dot{x}_f(t) = \sum\limits_{i=1}^{\ell} \mu_i(u_f(t))A_i x_f(t) + Bu_f(t) + Ef_a(t) \\ y_f(t) = Cx_f(t) + Dw(t) \end{cases} \tag{51}$$

where $x_f(t) \in \mathbb{R}^n$ is the faulty state vector, $u_f(t) \in \mathbb{R}^q$ is the fault tolerant control which will be conceived, $y_f(t) \in \mathbb{R}^m$ is the faulty output vector. $f_a(t)$ represents the actuator fault which is assumed to be bounded and $w(t)$ is the measurement noise. E and D are respectively the faults and the noise distribution matrices which are assumed to be known.

To estimate simultaneously the state x_f and the actuator fault f_a, a proportional integral observer is used, it is given by the following equations:

$$\begin{cases} \dot{\hat{x}}_f(t) = \sum\limits_{i=1}^{\ell} \mu_i(u_f(t))(A_i \hat{x}_f(t) + Bu_f(t) + K_i(y_f(t) - \hat{y}_f(t))) + E\hat{f}_a(t) \\ \dot{\hat{f}}_a(t) = \sum\limits_{i=1}^{\ell} \mu_i(u_f(t))L_i(y_f(t) - \hat{y}_f(t)) \\ \hat{y}_f(t) = C\hat{x}_f(t) \end{cases} \tag{52}$$

where $\hat{x}_f(t)$ is the estimated system state, $\hat{f}_a(t)$ is the estimated fault, $\hat{y}_f(t)$ is the estimated output, K_i are the proportional gains of the local observers and L_i are their integral gains to be computed.

The fuzzy fault tolerant control is conceived on the base of the following strategy.

$$u_f(t) = u(t) - S\hat{f}_a(t) + \sum_{i=1}^{\ell} \mu_i(u(t))G_i(x(t) - \hat{x}_f(t)) \tag{53}$$

where S and G_i are constant matrices with appropriate dimensions.

Let us define $\tilde{x}(t)$ the error between the states $x(t)$ and $x_f(t)$, $\tilde{x}_f(t)$ the estimation error of the state $x_f(t)$ and $\tilde{f}_a(t)$ the fault estimation error, these errors are given by :

$$\begin{aligned} \tilde{x}(t) &= x(t) - x_f(t) \\ \tilde{x}_f(t) &= x_f(t) - \hat{x}_f(t) \\ \tilde{f}_a(t) &= f_a(t) - \hat{f}_a(t) \end{aligned} \tag{54}$$

The matrix S is chosen verifying $E = BS$. The dynamics of $\tilde{x}(t)$ is given by:

$$\begin{aligned} \dot{\tilde{x}}(t) &= \dot{x}(t) - \dot{x}_f(t) \\ &= \sum_{i=1}^{\ell} \mu_i(u(t)) \sum_{j=1}^{\ell} \mu_j(u(t))((A_i - BG_j)\tilde{x}(t) \\ &-BG_j\tilde{x}_f) - E\tilde{f}_a(t) + \delta_1(t) \end{aligned} \tag{55}$$

with :

$$\delta_1(t) = \sum_{i=1}^{\ell}(\mu_i(u_f(t)) - \mu_i(u(t)))A_{ai}\tilde{x}_f(t) - D_a w(t)$$

The dynamic of $\tilde{x}_f(t)$ is given by:

$$\begin{aligned} \dot{\tilde{x}}_f(t) &= \dot{x}_f(t) - \dot{\hat{x}}_f(t) \\ &= \sum_{i=1}^{\ell} \mu_i(u(t))(A_i - K_iC)\tilde{x}_f(t) + E\tilde{f}(t) + \delta_2(t) \end{aligned} \tag{56}$$

with :

$$\delta_2(t) = \sum_{i=1}^{\ell}(\mu_i(u_f(t)) - \mu_i(u(t)))(A_i - K_iC)\tilde{x}_f(t) + Dw(t)$$

The dynamic of the fault error estimation is:

$$\begin{aligned} \dot{\tilde{f}}(t) &= \dot{f}(t) - \dot{\hat{f}}(t) \\ &= -\sum_{i=1}^{\ell} \mu_i(u(t))L_iC\tilde{x}_f(t) + \delta_3(t) \end{aligned} \tag{57}$$

with :

$$\delta_3(t) = \sum_{i=1}^{\ell}(\mu_i(u_f(t)) - \mu_i(u(t)))L_iC\tilde{x}_f(t) + Dw(t) + \dot{f}(t)$$

The Eqs. (55), (56) and (57) can be rewritten:

$$\dot{\varphi}(t) = A_m \varphi(t) + \varepsilon(t) \tag{58}$$

with :

$$\varphi(t) = \begin{bmatrix} \tilde{x}(t) \\ \tilde{x}_f(t) \\ \tilde{f}(t) \end{bmatrix} \quad \text{and} \quad \varepsilon(t) = \begin{bmatrix} \delta_1(t) \\ \delta_2(t) \\ \delta_3(t) \end{bmatrix}$$

$$\text{and } A_m = -\sum_{i=1}^{\ell} \mu_i(u(t)) \sum_{j=1}^{\ell} \mu_j(u(t)) A_{mij}$$

where

$$A_{mij} = \begin{bmatrix} A_i - BG_j & -BG_j & B \\ 0 & A_i - K_i C & B \\ 0 & L_i C & 0 \end{bmatrix}$$

Considering the Lyapunov function $V(t) = \varphi(t)^T P \varphi(t)$, the generalized error vector $\varphi(t)$ converges to zero if $\dot{V}(t) < 0 \Longrightarrow A_{mij}^T P + P A_{mij} < 0 \ \forall i, j \in \{1...\ell\}$.

The problem of robust state and faults estimation and of the fault tolerant control design is reduced to find the gains K_i and L_i of the proportional integral observer and the matrices G_i to ensure an asymptotic convergence of the generalized error vector $\varphi(t)$ toward zero if $\varepsilon(t) = 0$ and to ensure a bounded error in the case where $\varepsilon(t) \neq 0$, i.e.:

$$\begin{aligned} &\lim_{t \to \infty} \varphi(t) = 0 \text{ for } \ \varepsilon(t) = 0 \\ &\|\varphi(t)\|_{Q_\varphi} \leq \lambda \, \|\varepsilon(t)\|_{Q_\varepsilon} \ \text{ for } \varepsilon(t) \neq 0 \end{aligned} \tag{59}$$

where $\lambda > 0$ is the attenuation level. To satisfy theses constraints (59), it is sufficient to find a Lyapunov function $V(t)$ such that:

$$\dot{V}(t) + \varphi(t)^T Q_\varphi \varphi(t) - \lambda^2 \varepsilon(t)^T Q_\varepsilon \varepsilon(t) < 0 \tag{60}$$

where Q_φ and Q_ε are two positive definite matrices.

The inequality (60) can be written:

$$\begin{bmatrix} \varphi(t) \\ \varepsilon(t) \end{bmatrix}^T \Phi \begin{bmatrix} \varphi(t) \\ \varepsilon(t) \end{bmatrix} < 0 \tag{61}$$

where:

$$\Phi = \begin{bmatrix} A_m^T P + P A_m + Q_\varphi & P \\ P & -\lambda^2 Q_\varepsilon \end{bmatrix} \tag{62}$$

Choosing $Q_\varphi = Q_\varepsilon = I$ and assume that the Lyapunov matrix P has the form: $diag(I, P_2, P_3)$, the matrix Φ is written as following:

$$\Phi = \sum_{i=1}^{\ell} \mu_i(u(t)) \sum_{i=1}^{\ell} \mu_j(u(t))\Phi_{ij} \tag{63}$$

where:

$$\Phi_{ij} = \begin{bmatrix} \Phi_{11ij} & -BG_j & B & I & 0 & 0 \\ -G_j^T B^T & \Phi_{22i} & \Phi_{23i} & 0 & P_2 & 0 \\ B^T & \Phi_{32i} & I_3 & 0 & 0 & P_3 \\ I & 0 & 0 & \lambda_1 I_{01} & 0 & 0 \\ 0 & P_2 & 0 & 0 & \lambda_2 I_{02} & 0 \\ 0 & 0 & P_3 & 0 & 0 & \lambda_3 I_{03} \end{bmatrix} \tag{64}$$

with:

$$\begin{aligned} \Phi_{11ij} &= A_i - BG_j + A_i^T - G_j^T B^T + I_1 \\ \Phi_{22i} &= P_2 A_i - P_2 K_i C + A_i^T P_2 - C^T K_i^T P_2 + I_2 \\ \Phi_{23i} &= P_2 B + C^T L_i^T P_3 \\ \Phi_{32i} &= \Phi_{23i}^T \end{aligned}$$

$\Phi < 0$ if $\Phi_{ij} < 0 \forall i, j \in \{1...\ell\}$, the inequalities $\Phi_{ij} < 0$ are bilinear with regard to the variables K_i, L_i, P_2 and P_3, they can be linearised using the changes of variables : $U_{2i} = P_2 K_i$ and $U_{3i} = P_3 L_i$. The observer gains are computed after that using the equations:

$$\begin{aligned} K_i &= P_2^{-1} U_{2i} \\ L_i &= P_3^{-1} U_{3i} \end{aligned} \tag{65}$$

To summarize, we propose the following theorem describing the design of the observer and the fuzzy fault tolerant control:

Theorem 2 *The system (58) describing the evolution of the errors $\tilde{x}(t)$, $\tilde{x}_f(t)$ and $\tilde{f}(t)$ is stable if there exist symmetric definite positive matrices P_2 and P_3 and matrices U_{3i}, U_{2i} and G_j, $i, j \in \{1...\ell\}$ so that the LMI $\Phi_{ij} < 0$ are verified $\forall i, j \in \{1...\ell\}$ where :*

$$\Phi_{ij} = \begin{bmatrix} \Phi_{11ij} & -BG_j & B & I & 0 & 0 \\ -G_j^T B^T & \Phi_{22i} & \Phi_{23i} & 0 & P_2 & 0 \\ B^T & \Phi_{32i} & I_3 & 0 & 0 & P_3 \\ I & 0 & 0 & \lambda_1 I_{01} & 0 & 0 \\ 0 & P_2 & 0 & 0 & \lambda_2 I_{02} & 0 \\ 0 & 0 & P_3 & 0 & 0 & \lambda_3 I_{03} \end{bmatrix} \tag{66}$$

and:

$$\begin{aligned}\Phi_{11ij} &= A_i - BG_j + A_i^T - G_j^T B_a^T + I_1\\ \Phi_{22i} &= P_2A_i - P_2U_{2i} + A_i^T P_2 - C_a^T U_{2i}^T + I_2\\ \Phi_{23i} &= P_2B + C^T U_{3i}^T\\ \Phi_{32i} &= \Phi_{23i}^T\end{aligned} \tag{67}$$

The observer gains are obtained by: $L_i = P_3^{-1}U_{3i}$ *and* $K_i = P_2^{-1}U_{2i}$

The main advantage of the proposed method is to conceive simultaneously and separately the fuzzy fault tolerant control and the proportional integral observer.

6 Fuzzy Fault Tolerant Control Design for System with Sensor Faults

Consider the non linear system described by Takagi-Sugeno structure given by the Eq. (50):

A nonlinear Takagi-Sugeno model affected by sensor faults and measurement noise is gven by:

$$\begin{cases}\dot{x}_f(t) = \sum_{i=1}^{\ell} \mu_i(u_f(t))A_i x_f(t) + Bu_f(t)\\ y_f(t) = Cx_f(t) + Ff_s(t) + Dw(t)\end{cases} \tag{68}$$

where $x_f(t) \in \mathbb{R}^n$ is the faulty state vector, $u_f(t) \in \mathbb{R}^q$ is the fault tolerant control which will be conceived, $y_f(t) \in \mathbb{R}^m$ is the faulty output vector. $f_s(t)$ represents the sensor faults which is assumed to be bounded and $w(t)$ is the measurement noise. F and D are respectively the faults and the noise distribution matrices which are assumed to be constant and known.

The following states [13, 18, 27] are defined:

$$\begin{aligned}\dot{z}(t) &= \sum_{i=1}^{\ell} \mu_i(u(t))(-\bar{A}z(t) + \bar{A}Cx(t))\\ \dot{z}_f(t) &= \sum_{i=1}^{\ell} \mu_i(u_f(t))(-\bar{A}z(t) + \bar{A}Cx_f(t) + \bar{A}Ff_s(t) + \bar{A}Dw(t))\end{aligned} \tag{69}$$

where $-\bar{A}$ is a stable matrix with appropriate dimension.

Let us define the two following augmented states: $\boldsymbol{x}(t) = \begin{bmatrix} x(t)^T & z(t)^T \end{bmatrix}^T$ and $\boldsymbol{x}_f(t) = \begin{bmatrix} x_f^T(t) & z_f^T(t) \end{bmatrix}^T$, given by:

$$\begin{cases} \dot{\boldsymbol{x}}(t) = \sum\limits_{i=1}^{\ell} \mu_i(u(t)) A_{ai} \boldsymbol{x}(t) + B_a u(t) \\ \boldsymbol{y}(t) = C_a \boldsymbol{x}(t) \end{cases} \tag{70}$$

and

$$\begin{cases} \dot{\boldsymbol{x}}_f(t) = \sum\limits_{i=1}^{\ell} \mu_i(u_f(t)) A_{ai} \boldsymbol{x}_f(t) + B_a u_f(t) + F_a f_s(t) + D_a w(t) \\ \boldsymbol{y}_f(t) = C_a \boldsymbol{x}_f(t) \end{cases} \tag{71}$$

with:

$$A_{ai} = \begin{bmatrix} A_i & 0 \\ -\bar{A}C & -\bar{A} \end{bmatrix}, \quad F_a = \begin{bmatrix} 0 \\ \bar{A}F \end{bmatrix}, \quad D_a = \begin{bmatrix} 0 \\ \bar{A}D \end{bmatrix}$$

$$B_a = \begin{bmatrix} B \\ 0 \end{bmatrix} \quad \text{and} \quad C_a = \begin{bmatrix} 0 & I \end{bmatrix}$$

A proportional integral observer able to estimate the augmented state $\boldsymbol{x}_f(t)$ and the sensor fault f_s given by the following equations is used:

$$\begin{cases} \dot{\hat{\boldsymbol{x}}}_f(t) = \sum\limits_{i=1}^{\ell} \mu_i(u_f(t))(A_{ai} \hat{\boldsymbol{x}}_f(t) + K_i \tilde{\boldsymbol{y}}_f(t)) + F_a \hat{f}(t) + B_a u_f(t) \\ \dot{\hat{f}}(t) = \sum\limits_{i=1}^{\ell} \mu_i(u_f(t)) L_i \tilde{\boldsymbol{y}}_f(t) \\ \hat{\boldsymbol{y}}_f(t) = C_a \hat{\boldsymbol{x}}_f(t) \end{cases} \tag{72}$$

where $\hat{\boldsymbol{x}}_f(t)$ is the estimated system state, $\hat{f}_s(t)$ represents the estimated sensor fault, $\hat{\boldsymbol{y}}_f(t)$ is the estimated output, K_i are the proportional gains of the local observers and L_i are their integral gains to be computed and $\tilde{\boldsymbol{y}}_f(t) = \boldsymbol{y}_f(t) - \hat{\boldsymbol{y}}_f(t)$.

The fuzzy fault tolerant control is conceived on the base of the following strategy.

$$u_f(t) = u(t) - S\hat{f}_s(t) + \sum_{i=1}^{\ell} \mu_i(u(t)) G_i(\boldsymbol{x}(t) - \hat{\boldsymbol{x}}_f(t)) \tag{73}$$

where S and G_i are constant matrices with appropriate dimensions. Let us define $\tilde{\boldsymbol{x}}(t)$ the error between the states $\boldsymbol{x}(t)$ and $\boldsymbol{x}_f(t)$, $\tilde{\boldsymbol{x}}_f(t)$ the estimation error of the faulty state $\boldsymbol{x}_f(t)$ and $\tilde{f}(t)$ the sensor fault estimation error. These errors are written as following:

$$\begin{aligned} \tilde{\boldsymbol{x}}(t) &= \boldsymbol{x}(t) - \boldsymbol{x}_f(t) \\ \tilde{\boldsymbol{x}}_f(t) &= \boldsymbol{x}_f(t) - \hat{\boldsymbol{x}}_f(t) \\ \tilde{f}(t) &= f(t) - \hat{f}(t) \end{aligned} \tag{74}$$

The matrix S is chosen verifying $F_a = B_a S$. The dynamics of $\tilde{x}(t)$ is given by:

$$\begin{aligned}\dot{\tilde{x}}(t) &= \dot{x}(t) - \dot{x}_f(t) \\ &= \sum_{i=1}^{\ell} \mu_i(u(t)) \sum_{j=1}^{\ell} \mu_j(u(t))(A_{ai} - B_a G_j)\tilde{x}(t) - F_a \tilde{f}(t) - B_a G \tilde{x}_f(t) + \Delta_1(t)\end{aligned} \tag{75}$$

with :

$$\Delta_1(t) = \sum_{i=1}^{\ell} (\mu_i(u_f(t)) - \mu_i u(t)) A_{ai} \tilde{x}_f(t) - D_a w(t)$$

The dynamic of $\tilde{x}_f(t)$ is written as:

$$\begin{aligned}\dot{\tilde{x}}_f(t) &= \dot{x}_f(t) - \dot{\hat{x}}_f(t) \\ &= \sum_{i=1}^{\ell} \mu_i(u(t))(A_{ai} - K_i C_a)\tilde{x}_f(t) + E_a \tilde{f}(t) + \Delta_2(t)\end{aligned} \tag{76}$$

with :

$$\Delta_2(t) = \sum_{i=1}^{\ell} \mu_i((u_f(t)) - \mu_i u(t))(A_{ai} - K_i C_a)\tilde{x}_f(t) + D_a w(t)$$

The dynamic of the sensor fault estimation error is:

$$\begin{aligned}\dot{\tilde{f}}(t) &= \dot{f}(t) - \dot{\hat{f}}(t) \\ &= -\sum_{i=1}^{\ell} \mu_i(u(t)) L_i C_a \tilde{x}_f(t) + \Delta_3(t)\end{aligned} \tag{77}$$

with :

$$\Delta_3(t) = \sum_{i=1}^{\ell} \mu_i(u_f(t) - \mu_i u(t)) L_i C_a \tilde{x}_f(t) + D_a w(t) + \dot{f}(t)$$

The equations (75), (76) and (77) can be rewritten in the following generalized form:

$$\dot{\varphi}(t) = A_m \varphi(t) + \varepsilon(t) \tag{78}$$

with :

$$\varphi(t) = \begin{bmatrix} \tilde{x}(t) \\ \tilde{x}_f(t) \\ \tilde{f}(t) \end{bmatrix} \quad \text{and} \quad \varepsilon(t) = \begin{bmatrix} \Delta_1(t) \\ \Delta_2(t) \\ \Delta_3(t) \end{bmatrix}$$

$$\text{and } A_m = \sum_{i=1}^{\ell} \mu_i(u(t)) \sum_{j=1}^{\ell} \mu_j(u(t)) A_{mij}$$

where

$$A_{mij} = \begin{bmatrix} A_{ai} - B_a G_j & -B_a G_j & B_a \\ 0 & A_{ai} - K_i C_a & B_a \\ 0 & L_i C_a & 0 \end{bmatrix}$$

Considering the Lyapunov function $V(t) = \varphi(t)^T P \varphi(t)$, the generalized error vector $\varphi(t)$ converges to zero if $\dot{V}(t) < 0 \Longrightarrow A_{mij}^T P + P A_{mij} < 0 \forall i, j \in \{1...\ell\}$. The problem of robust state and faults estimation and of the fault tolerant control design is reduced to find the gains K_i and L_i of the observer and the matrices G_j to ensure an asymptotic convergence of the generalized error vector $\varphi(t)$ toward zero if $\varepsilon(t) = 0$ and to ensure a bounded error if $\varepsilon(t) \neq 0$, i.e.:

$$\begin{array}{ll} \lim_{t \to \infty} \varphi(t) = 0 & \text{for} \varepsilon(t) = 0 \\ \|\varphi(t)\|_{Q_\varphi} \leq \lambda \|\varepsilon(t)\|_{Q_\varepsilon} & \text{for} \varepsilon(t) \neq 0 \end{array} \tag{79}$$

where $\lambda > 0$ is the attenuation level. To satisfy the constraints (79), it is sufficient to find a Lyapunov function $V(t)$ such that:

$$\dot{V}(t) + \varphi(t)^T Q_\varphi \varphi(t) - \lambda^2 \varepsilon(t)^T Q_\varepsilon \varepsilon(t) < 0 \tag{80}$$

where Q_φ and Q_ε are two positive definite matrices.

The inequality (80) can be written as:

$$\begin{bmatrix} \varphi(t) \\ \varepsilon(t) \end{bmatrix}^T \Phi \begin{bmatrix} \varphi(t) \\ \varepsilon(t) \end{bmatrix} < 0 \tag{81}$$

where:

$$\Phi = \begin{bmatrix} A_m^T P + P A_m + Q_\varphi & P \\ P & -\lambda^2 Q_\varepsilon \end{bmatrix} \tag{82}$$

Choosing $Q_\varphi = Q_\varepsilon = I$ and assume that the Lyapunov matrix P has the form: $diag(I, P_2, P_3)$, the matrix Φ is written :

$$\Phi = \sum_{i=1}^{\ell} \mu_i(u(t)) \sum_{j=1}^{\ell} \mu_j(u(t)) \Phi_{ij} \tag{83}$$

where:

$$\Phi_{ij} = \begin{bmatrix} \Phi_{11ij} & -B_a G_j & B_a & I & 0 & 0 \\ -G_j^T B_a^T & \Phi_{22i} & \Phi_{23i} & 0 & P_2 & 0 \\ B_a^T & \Phi_{32i} & I_3 & 0 & 0 & P_3 \\ I & 0 & 0 & \lambda_1 I_{01} & 0 & 0 \\ 0 & P_2 & 0 & 0 & \lambda_2 I_{02} & 0 \\ 0 & 0 & P_3 & 0 & 0 & \lambda_3 I_{03} \end{bmatrix} \tag{84}$$

with:

$$\begin{aligned} \Phi_{11ij} &= A_{ai} - B_a G_j + A_{ai}^T - G_j^T B_a^T + I_1 \\ \Phi_{22i} &= P_2 A_{ai} - P_2 K_i C_a + A_{ai}^T P_2 - C_a^T K_i^T P_2 + I_2 \\ \Phi_{23i} &= P_2 B_a + C_a^T L_i^T P_3 \\ \Phi_{32i} &= \Phi_{23i}^T \end{aligned}$$

$\Phi < 0$ if $\Phi_{ij} < 0 \forall i, j \in \{1 \ldots \ell\}$, the inequalities $\Phi_{ij} < 0$ are bilinear, they can be linearised using the changes of variables : $U_{2i} = P_2 K_i$ and $U_{3i} = P_3 L_i$. The observer gains are then computed using the equations:

$$\begin{aligned} K_i &= P_2^{-1} U_{2i} \\ L_i &= P_3^{-1} U_{3i} \end{aligned} \tag{85}$$

To summarize, we propose the following theorem:

Theorem 3 *The system (78) describing the evolution of the errors $\tilde{x}(t)$, $\tilde{x}_f(t)$ and $\tilde{f}(t)$ is stable if there exist symmetric definite positive matrices P_2 and P_3 and matrices U_{3i}, U_{2i} and G_j, $i, j \in \{1 \ldots \ell\}$ so that the LMI $\Phi_i < 0$ are verified $\forall i \in \{1 \ldots \ell\}$ where :*

$$\Phi_{ij} = \begin{bmatrix} \Phi_{11ij} & -B_a G_j & B_a & I & 0 & 0 \\ -G_j^T B_a^T & \Phi_{22i} & \Phi_{23i} & 0 & P_2 & 0 \\ B_a^T & \Phi_{32i} & I_3 & 0 & 0 & P_3 \\ I & 0 & 0 & \lambda_1 I_{01} & 0 & 0 \\ 0 & P_2 & 0 & 0 & \lambda_2 I_{02} & 0 \\ 0 & 0 & P_3 & 0 & 0 & \lambda_3 I_{03} \end{bmatrix} \tag{86}$$

and:

$$\begin{aligned} \Phi_{11ij} &= A_{ai} - B_a G_j + A_{ai}^T - G_j^T B_a^T + I_1 \\ \Phi_{22i} &= P_2 A_{ai} - P_2 U_{2i} + A_{ai}^T P_2 - C_a^T U_{2i}^T + I_2 \\ \Phi_{23i} &= P_2 B_a + C_a^T U_{3i}^T \\ \Phi_{32i} &= \Phi_{23i}^T \end{aligned} \tag{87}$$

The observer gains are obtained by: $L_i = P_3^{-1} U_{3i}$ and $K_i = P_2^{-1} U_{2i}$

The advantage of this method is to conceive simultaneously the proportional integral observer and the fuzzy fault tolerant control. A mathematical transformation based on the use of the new states z and z_f permits to conceive augmented states in where the initial sensor fault appears as an actuator fault. The fuzzy fault tolerant control is conceived with regard to the augmented states $\boldsymbol{x}(t)$ and $\boldsymbol{x}_f(t)$

7 Fuzzy Fault Tolerant Control Design for System with Actuator and Sensor Faults

Consider the non linear system described by Takagi-Sugeno structure given by the Eq. (50). A nonlinear Takagi-Sugeno model affected by actuator and sensor faults and measurement noise is given by the following equation:

$$\begin{cases} \dot{x}_f(t) = \sum_{i=1}^{\ell} \mu_i(u_f(t))A_i x_f(t) + Bu_f(t) + Ef_a(t) \\ y_f(t) = Cx_f(t) + Ff_s(t) + Dw(t) \end{cases} \tag{88}$$

where $x_f(t) \in \mathbb{R}^n$ is the faulty state vector, $u_f(t) \in \mathbb{R}^q$ is the fuzzy fault tolerant control which will be conceived, $y_f(t) \in \mathbb{R}^m$ is the faulty output vector. $f_a(t)$ and $f_s(t)$ are respectively the actuator and the sensor faults which are assumed to be bounded and $w(t)$ is the measurement noise. E, F and D are respectively the faults and the noise distribution matrices which are assumed to be known.

Using the states z and z_f defined in (69), the two augmented states $\boldsymbol{x}(t) = \begin{bmatrix} x(t)^T & z(t)^T \end{bmatrix}^T$ and $\boldsymbol{x}_f(t) = \begin{bmatrix} x_f^T(t) & z_f^T(t) \end{bmatrix}^T$, can be written:

$$\begin{cases} \dot{\boldsymbol{x}}(t) = \sum_{i=1}^{\ell} \mu_i(u(t))A_{ai}\boldsymbol{x}(t) + B_a u(t) \\ \boldsymbol{y}(t) = C_a \boldsymbol{x}(t) \end{cases} \tag{89}$$

and

$$\begin{cases} \dot{\boldsymbol{x}}_f(t) = \sum_{i=1}^{\ell} \mu_i(u_f(t))A_{ai}\boldsymbol{x}_f(t) + B_a u_f(t) + E_a f(t) + D_a w(t) \\ \boldsymbol{y}_f(t) = C_a \boldsymbol{x}_f(t) \end{cases} \tag{90}$$

with:

$$A_{ai} = \begin{bmatrix} A_i & 0 \\ -\bar{A}C & -\bar{A} \end{bmatrix}, \; E_a = \begin{bmatrix} E & 0 \\ 0 & \bar{A}F \end{bmatrix}, \; f = \begin{bmatrix} f_a \\ f_s \end{bmatrix},$$

$$B_a = \begin{bmatrix} B \\ 0 \end{bmatrix}, \; D_a = \begin{bmatrix} 0 \\ \bar{A}D \end{bmatrix} \text{ and } C_a = \begin{bmatrix} 0 & I \end{bmatrix}$$

In order to estimate the augmented state $\boldsymbol{x}_f$ and the generalized fault f, a proportional integral observer given by the following equations is used:

$$\begin{cases} \dot{\hat{\boldsymbol{x}}}_f(t) = \sum_{i=1}^{\ell} \mu_i(u_f(t))(A_{ai}\hat{\boldsymbol{x}}_f(t) + K_i\tilde{\boldsymbol{y}}_f(t)) + B_a u_f(t) + E_a\hat{f}(t) \\ \dot{\hat{f}}(t) = \sum_{i=1}^{\ell} \mu_i(u_f(t))L_i\tilde{\boldsymbol{y}}_f(t) \\ \hat{\boldsymbol{y}}_f(t) = C_a\hat{\boldsymbol{x}}_f(t) \end{cases} \tag{91}$$

where $\hat{\boldsymbol{x}}_f(t)$ is the estimated faulty state, $\hat{f}(t)$ represents the estimated generalized fault, $\hat{\boldsymbol{y}}_f(t)$ is the estimated system output, K_i are the proportional gains of the local observers and L_i are their integral gains to be computed and $\tilde{\boldsymbol{y}}_f(t) = \boldsymbol{y}_f(t) - \hat{\boldsymbol{y}}_f(t)$.

The fault tolerant control is conceived on the base of the strategy described by the following expression.

$$u_f(t) = u(t) - S\hat{f}(t) + \sum_{i=1}^{\ell} \mu_i(u(t))G_i(\boldsymbol{x}(t) - \hat{\boldsymbol{x}}_f(t)) \tag{92}$$

where S and G_i are constant matrices with appropriate dimensions.

The following errors are used :

$$\begin{aligned} \tilde{\boldsymbol{x}}(t) &= \boldsymbol{x}(t) - \boldsymbol{x}_f(t) \\ \tilde{\boldsymbol{x}}_f(t) &= \boldsymbol{x}_f(t) - \hat{\boldsymbol{x}}_f(t) \\ \tilde{f}(t) &= f(t) - \hat{f}(t) \end{aligned} \tag{93}$$

The matrix S is chosen verifying $E_a = B_a S$. The dynamics of $\tilde{\boldsymbol{x}}(t)$ is given by:

$$\begin{aligned} \dot{\tilde{\boldsymbol{x}}}(t) &= \dot{\boldsymbol{x}}(t) - \dot{\boldsymbol{x}}_f(t) \\ &= \sum_{i=1}^{\ell} \mu_i(u(t)) \sum_{j=1}^{\ell} \mu_j(u(t))((A_{ai} - B_a G_j)\tilde{\boldsymbol{x}}(t) - B_a G_j\tilde{\boldsymbol{x}}_f(t)) - E_a\tilde{f}(t) + \boldsymbol{\Delta}_1(t) \end{aligned} \tag{94}$$

with:

$$\boldsymbol{\Delta}_1(t) = \sum_{i=1}^{\ell}(\mu_i(u_f(t)) - \mu_i(u(t)))A_{ai}\tilde{\boldsymbol{x}}_f(t) - D_a w(t)$$

The dynamic of $\tilde{\boldsymbol{x}}_f(t)$ can be written:

$$\begin{aligned} \dot{\tilde{\boldsymbol{x}}}_f(t) &= \dot{\boldsymbol{x}}_f(t) - \dot{\boldsymbol{x}}_f(t) \\ &= \sum_{i=1}^{\ell} \mu_i(u(t))(A_{ai} - K_i C_a)\tilde{\boldsymbol{x}}_f(t) + E_a\tilde{f}(t) + \boldsymbol{\Delta}_2(t) \end{aligned} \tag{95}$$

with :

$$\boldsymbol{\Delta}_2(t) = \sum_{i=1}^{\ell}(\mu_i(u_f(t)) - \mu_i(u(t)))(A_{ai} - K_i C_a)\tilde{\boldsymbol{x}}_f(t) + D_a w(t)$$

The dynamic of the fault error estimation is:

$$\begin{aligned}\dot{\tilde{f}}(t) &= \dot{f}(t) - \dot{\hat{f}}(t) \\ &= -\sum_{i=1}^{\ell} \mu_i(u(t)) L_i C_a \tilde{\boldsymbol{x}}_f(t) + \boldsymbol{\Delta}_3(t)\end{aligned} \tag{96}$$

with :

$$\boldsymbol{\Delta}_3(t) = \sum_{i=1}^{\ell}(\mu_i(u_f(t)) - \mu_i(u(t)))L_i C_a \tilde{\boldsymbol{x}}_f(t) + D_a w(t) + \dot{f}(t)$$

The Eqs. (94), (95) and (96) can be rewritten:

$$\dot{\varphi}(t) = A_m \varphi(t) + \varepsilon(t) \tag{97}$$

with :

$$\varphi(t) = \begin{bmatrix} \tilde{\boldsymbol{x}}(t) \\ \tilde{\boldsymbol{x}}_f(t) \\ \tilde{f}(t) \end{bmatrix} \quad \text{and} \quad \varepsilon(t) = \begin{bmatrix} \boldsymbol{\Delta}_1(t) \\ \boldsymbol{\Delta}_2(t) \\ \boldsymbol{\Delta}_3(t) \end{bmatrix}$$

$$\text{and } A_m = -\sum_{i=1}^{\ell} \mu_i(u(t)) \sum_{j=1}^{\ell} \mu_j(u(t)) A_{mij}$$

where

$$A_{mij} = \begin{bmatrix} A_{ai} - B_a G_j & -B_a G_j & B_a \\ 0 & A_{ai} - K_i C_a & B_a \\ 0 & L_i C_a & 0 \end{bmatrix}$$

Considering the Lyapunov function $V(t) = \varphi(t)^T P \varphi(t)$, the generalized error vector $\varphi(t)$ converges to zero if $\dot{V}(t) < 0 \Longrightarrow A_{mij}^T P + P A_{mij} < 0 \forall i, j \in \{1...\ell\}$.

The problem of robust state and faults estimation and of the fault tolerant control design is reduced to find the gains K_i and L_i of the observer and the matrices G_j to ensure an asymptotic convergence of the generalized error vector $\varphi(t)$ toward zero if $\varepsilon(t) = 0$ and to ensure a bounded error in the case where $\varepsilon(t) \neq 0$, i.e.:

$$\begin{aligned} &\lim_{t\to\infty} \varphi(t) = 0 && \text{for } \varepsilon(t) = 0 \\ &\|\varphi(t)\|_{Q_\varphi} \leq \lambda \|\varepsilon(t)\|_{Q_\varepsilon} && \text{for } \varepsilon(t) \neq 0 \end{aligned} \tag{98}$$

where $\lambda > 0$ is the attenuation level. The constraints (98), are satisfied if we to find a Lyapunov function $V(t)$ such that:

$$\dot{V}(t) + \varphi(t)^T Q_\varphi \varphi(t) - \lambda^2 \varepsilon(t)^T Q_\varepsilon \varepsilon(t) < 0 \tag{99}$$

where Q_φ and Q_ε are two positive definite matrices.

The inequality (99) can be written:

$$\begin{bmatrix} \varphi(t) \\ \varepsilon(t) \end{bmatrix}^T \Phi \begin{bmatrix} \varphi(t) \\ \varepsilon(t) \end{bmatrix} < 0 \tag{100}$$

where:

$$\Phi = \begin{bmatrix} A_m^T P + P A_m + Q_\varphi & P \\ P & -\lambda^2 Q_\varepsilon \end{bmatrix} \tag{101}$$

Choosing $Q_\varphi = Q_\varepsilon = I$ and assume that the Lyapunov matrix P has the form: $diag(I, P_2, P_3)$, the matrix Φ is written :

$$\Phi = \sum_{i=1}^{\ell} \mu_i(u(t)) \sum_{i=1}^{\ell} \mu_i(u(t)) \Phi_{ij} \tag{102}$$

where:

$$\Phi_{ij} = \begin{bmatrix} \Phi_{11ij} & -B_a G_j & B_a & I & 0 & 0 \\ -G_j^T B_a^T & \Phi_{22i} & \Phi_{23i} & 0 & P_2 & 0 \\ B_a^T & \Phi_{32i} & I_3 & 0 & 0 & P_3 \\ I & 0 & 0 & \lambda_1 I_{01} & 0 & 0 \\ 0 & P_2 & 0 & 0 & \lambda_2 I_{02} & 0 \\ 0 & 0 & P_3 & 0 & 0 & \lambda_3 I_{03} \end{bmatrix} \tag{103}$$

with:

$$\begin{aligned} \Phi_{11ij} &= A_{ai} - B_a G_j + A_{ai}^T - G_j^T B_a^T + I_1 \\ \Phi_{22i} &= P_2 A_{ai} - P_2 K_i C_a + A_{ai}^T P_2 - C_a^T K_i^T P_2 + I_2 \\ \Phi_{23i} &= P_2 B_a + C_a^T L_i^T P_3 \\ \Phi_{32i} &= \Phi_{23i}^T \end{aligned}$$

$\Phi < 0$ if $\Phi_{ij} < 0 \forall i, j \in \{1...\ell\}$, the inequalities $\Phi_i < 0$ are bilinear, they can be linearised using the changes of variables : $U_{2i} = P_2 K_i$ and $U_{3i} = P_3 L_i$. The observer gains are then computed using the equations:

$$\begin{aligned} K_i &= P_2^{-1} U_{2i} \\ L_i &= P_3^{-1} U_{3i} \end{aligned} \tag{104}$$

To summarize, we propose the following theorem:

Theorem 4 *The system (97) describing the evolution of the errors $\tilde{x}(t), \tilde{x}_f(t)$ and $\tilde{f}(t)$ is stable if there exist symmetric definite positive matrices P_2 and P_3 and matrices U_{3i}, U_{2i} and $G_j, i, j \in \{1...\ell\}$ so that the LMI $\Phi_{ij} < 0$ are verified $\forall i, j \in \{1...\ell\}$ where :*

$$\Phi_{ij} = \begin{bmatrix} \Phi_{11ij} & -B_a G_j & B_a & I & 0 & 0 \\ -G_j^T B_a^T & \Phi_{22i} & \Phi_{23i} & 0 & P_2 & 0 \\ B_a^T & \Phi_{32i} & I_3 & 0 & 0 & P_3 \\ I & 0 & 0 & \lambda_1 I_{01} & 0 & 0 \\ 0 & P_2 & 0 & 0 & \lambda_2 I_{02} & 0 \\ 0 & 0 & P_3 & 0 & 0 & \lambda_3 I_{03} \end{bmatrix} \tag{105}$$

and:

$$\begin{aligned} \Phi_{11ij} &= A_{ai} - B_a G_j + A_{ai}^T - G_j^T B_a^T + I_1 \\ \Phi_{22i} &= P_2 A_{ai} - P_2 U_{2i} + A_{ai}^T P_2 - C_a^T U_{2i}^T + I_2 \\ \Phi_{23i} &= P_2 B_a + C_a^T U_{3i}^T \\ \Phi_{32i} &= \Phi_{23i}^T \end{aligned} \tag{106}$$

The observer gains are obtained by: $L_i = P_3^{-1} U_{3i}$ *and* $K_i = P_2^{-1} U_{2i}$

8 Example of Simulation

The main objective of this part is to show the robustness of the proposed methods by its application to a hydraulic process made up of three tanks [3]. The system is supposed affected simultaneously by sensor and actuator faults. The considered system for this application is described and modeled in [3]. The non linear model is given by the Eq. (107) [3]. where:

- $Q_1(t)$ and $a_2(t)$ are the flow rates.
- ρ is the tanks section.
- S_n is the cylindrical pipes sections.
- α_l, α_2 and α_3 are constants.
- $Qf/fi(t), ie1.....3$ denote the additional mass flows.
- x_1, x_2 and x_3 are the water level with $x_1 > x_3 > x_2$.

$$\begin{cases} \rho\frac{dx_1(t)}{dt} = -\alpha_1 S_n(2g(x_1(t) - x_3(t)))^{1/2} + Q_1(t) + Qf_1\bar{u}(t) \\ \rho\frac{dx_2(t)}{dt} = -\alpha_3 S_n(2g(x_3(t) - x_2(t)))^{1/2} - \alpha_2 S_n(2gx_2(t))^{1/2} + Q_2(t) + Qf_2\bar{u}(t) \\ \rho\frac{dx_3(t)}{dt} = -\alpha_1 S_n(2g(x_1(t) - x_3(t)))^{1/2} - \alpha_3 S_n(2g(x_3(t) - x_2(t)))^{1/2} + Qf_3\bar{u}(t) \end{cases} \tag{107}$$

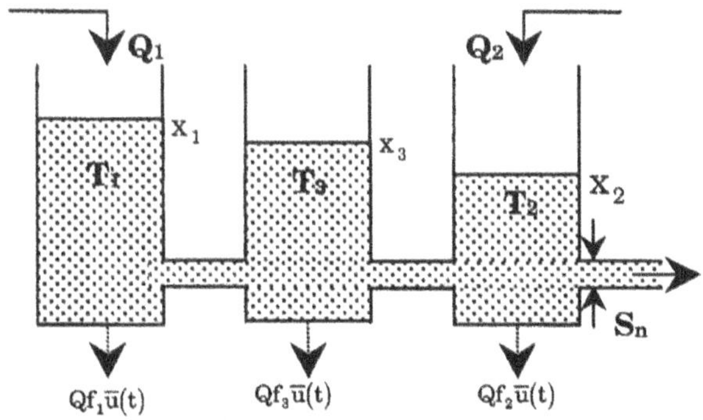

Fig. 1 Three tanks system

The process is given in Fig. 1. More details about the used process can be found in [3].

The multiple model, based on the Takagi-Sugeno framework, is used for the simulation. This model is given in Eq. 108, with $\xi(t) = u(t)$.

$$\dot{x}(t) = \sum_{i=1}^{\ell} \mu_i(\xi(t))(A_i x(t) + B_i u(t) + F\bar{u}(t) + d_i) \tag{108a}$$

$$y(t) = Cx(t) + Dw(t) \tag{108b}$$

The matrices A_i, Bi and d_i are calculated by linearizing the initial system (107) around four points chosen in the operation range of the system. Four local models have been selected in an heuristic way. That number guarantees a good approximation of the state of the real system by the multiple model [3]. The following numerical values were obtained:

$$A_1 = \begin{bmatrix} -0.0109 & 0 & 0.0109 \\ 0 & -0.0206 & 0.0106 \\ 0.0109 & 0.0106 & -0.0215 \end{bmatrix}, \quad d_1 = 10^{-3} * \begin{bmatrix} -2.86 \\ -0.38 \\ 0.11 \end{bmatrix}$$

$$A_2 = \begin{bmatrix} -0.0110 & 0 & 0.0110 \\ 0 & -0.0205 & 0.0104 \\ 0.0110 & 0.0104 & -0.0215 \end{bmatrix}, \quad d_2 = 10^{-3} * \begin{bmatrix} -2.86 \\ -0.34 \\ 0.038 \end{bmatrix}$$

$$A_3 = \begin{bmatrix} -0.0084 & 0 & 0.0084 \\ 0 & -0.0206 & 0.0095 \\ 0.0084 & 0.0095 & -0.0180 \end{bmatrix}, \quad d_3 = 10^{-3} * \begin{bmatrix} -3.7 \\ -0.14 \\ 0.69 \end{bmatrix}$$

$$A_4 = \begin{bmatrix} -0.0085 & 0 & 0.0085 \\ 0 & -0.0205 & 0.0095 \\ 0.0085 & 0.0095 & -0.0180 \end{bmatrix}, \quad d_4 = 10^{-3} * \begin{bmatrix} -3.67 \\ -0.18 \\ 0.62 \end{bmatrix}$$

$$B_i = \frac{1}{A} * \begin{bmatrix} 1 & 0 \\ 0 & 1 \\ 0 & 0 \end{bmatrix} \quad \text{and} \quad C = \begin{bmatrix} 1 & 1 & 1 \\ 1 & 0 & 0 \\ 0 & 1 & 0 \end{bmatrix}$$

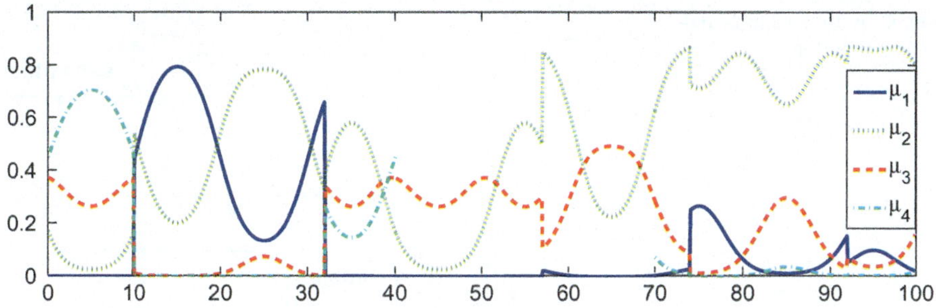

Fig. 2 Activation functions

Fig. 3 System input

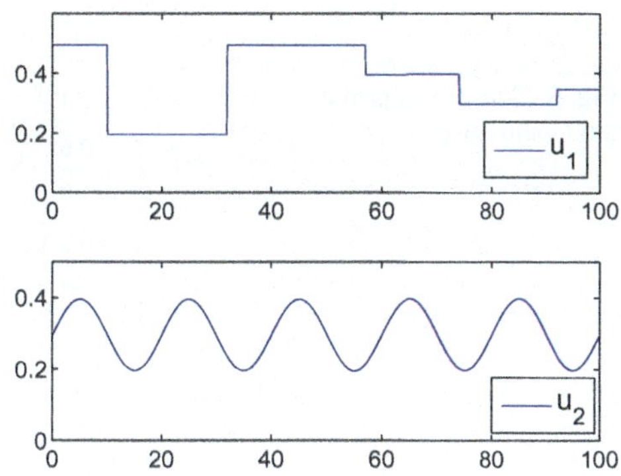

The activation functions $\mu_i(u(t))$ are shown in the Fig. 2.

The system input $u(t)$ is chosen with two components $u(t) = \left[u_1(t)\ u_2(t) \right]^T$ where u_1 is a signal varying between 0 and 0.5 and $u_2(t) = 0.3 + 0.1 * \sin(\pi t)$. The system input is shown in Fig. 3

The actuator fault $f_a(t) = \left[f_{a1}(t)\ f_{a2}(t) \right]^T$ is defined as:

$$f_{a1} = \begin{cases} 0.4 * \sin(\pi t), & 150s < t < 750s \\ 0, & \text{Otherwise} \end{cases} \quad \text{and} \quad f_{a2} = \begin{cases} 0.3, & 200s < t < 700s \\ 0.5, & 700s < t < 1000s \\ 0, & \text{Otherwise} \end{cases}$$

The sensor fault $f_s(t)$ is: $f_s(t) = \left[f_{s1}(t)\ f_{s2}(t) \right]^T$ with:

$$f_{s1} = \begin{cases} 0.4 * \sin(\pi t), & 100s < t < 300s \text{ and } 550s < t < 750s \\ 0.5, & 300s < t < 550s \\ 0, & \text{Otherwise} \end{cases} \quad \text{and}$$

$$f_{s2} = \begin{cases} 0.5, & 100s < t < 200s \text{ and } 700s < t < 800s \\ 0.5 + \sin(1.2 * \pi t), & 200s < t < 700s \\ 0, & \text{Otherwise} \end{cases}$$

Matrix $\bar{A}$ is chosen as: $\bar{A} = 10 * I$.

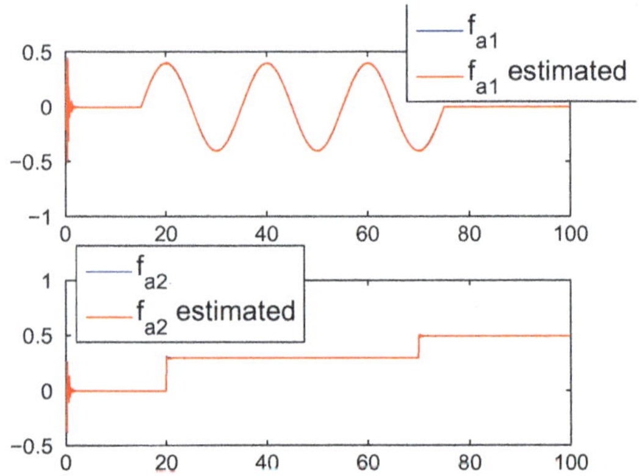

Fig. 4 Actuator faults and their estimation

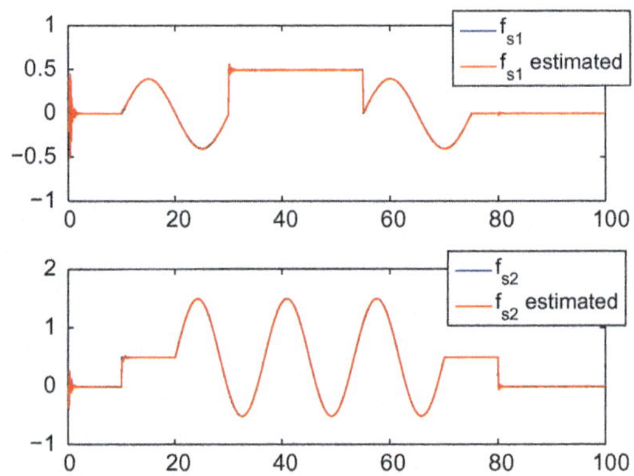

Fig. 5 Sensor faults and their estimation

Figure 4 visualizes the two actuator faults and their estimations. In Fig. 5 the two sensor faults and their estimations are represented. The state error estimation is visualized in Fig. 6.

Figures (4) and (5) show that the proposed proportional observer allows estimating well the actuator and the sensor faults even in the case of time-varying faults. Figure (6) shows also that this proposed observer allows estimating well the faulty system state. The effect of the measurement noise is minimized using a $\mathcal{L}_2$ approach.

The obtained results show the effectiveness of the proposed proportional integral observer. Figures 4, 5, 6 show the effectiveness of the proposed proportional integral observer to estimate state and faults estimation with high performances. The proposed observer gives a good estimation even in the case of time-varying faults and in a very short time.

Figure 7 shows the error between the initial state x which is taken as a reference and the faulty system state x_f. The fuzzy fault tolerant control is presented in Fig. 8. Figure 7 shows that the error between the state x and the faulty state x_f converges to zero. In other words, this result means that the proposed fault tolerant control take action rapidly on the faulty system state x_f to make its behavior similar

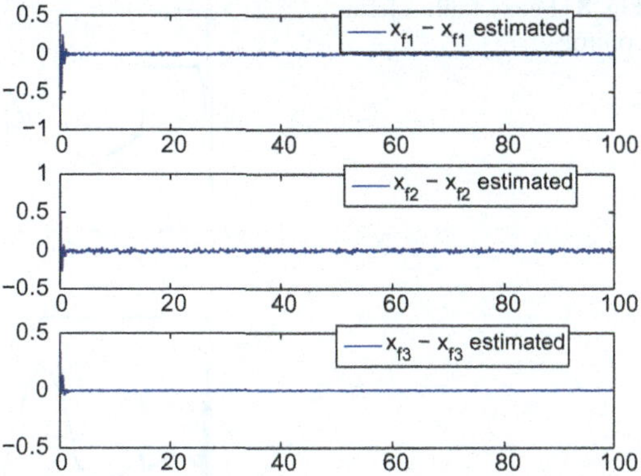

Fig. 6 State estimation error

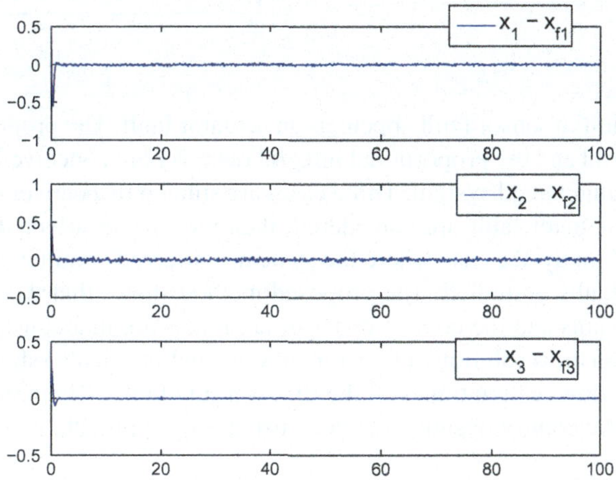

Fig. 7 Error between x and x_f

to the behavior of the system without fault described by the state x. The proposed fuzzy fault tolerant control is shown in Eq. 53. This figure shows that the fault tolerant control changes in real time to take into account the sensor and actuators faults variations

9 Conclusion

This chapter presents a method of a fuzzy fault tolerant control design based on the principle of Takagi-Sugeno systems. The proposed fault tolerant control depends on the state estimation error and the fault estimation. To make this estimation, a proportional integral observer allowing estimating simultaneously the system state and the fault is used. To consider the case where the system is affected by sensor faults, a mathematical transformation is used to conceive an augmented system in which the

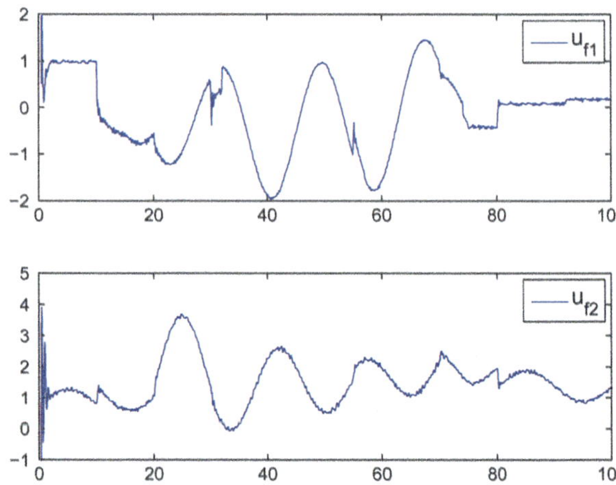

Fig. 8 Fuzzy fault tolerant control

initial sensor fault appear as an actuator fault. The proposed fuzzy fault tolerant control and the proportional integral observer are conceived by considering the obtained augmented system. Three cases are studied respectively. Firstly, systems affected by actuator fault are considered, then the system affected by sensor fault are treated. Finally, the case where the system is affected simultaneously by sensor and actuator faults is studied. The proposed method shows that the computation of the observer gains and the control matrices are made simultaneously. This computation is based on the resolution of linear matrix inequalities. It also shown that the proposed method allows estimating well the time-varying faults. The noise effect on the estimation or the control design is reduced using a $\mathcal{L}_2$ approach.

References

1. Acevedo-Valle, J.M., Puig, V., Tornil-Sin, S., Witczak, M., Rotondo, D., Control, Predictive Fault Tolerant., for LPV systems using model reference, Preprints. In: 9th IFAC Symposium on Fault Detection, Supervision and Safety of Technical Processes, Arts et Métiers ParisTech. France, Paris (2015)
2. Ait Daraou K., Soulami J., El Assoudi A., El Yaagoubi E.H.: State and fault observer design for discrete-time Takagi-Sugeno descriptor models. In: 1st International Conference on Innovative Research in Applied Science, Engineering and Technology (IRASET) (2020)
3. Akhenak A., Chadli M., Maquin D., Ragot J.: state estimation via multiple obserer. The three tank system. In: 5th IFAC Symposium on Fault Detection, Supervision and Safety for Technical Processes, Safeprocess'2003, Washington, D.C., USA (2003)
4. Asemani, M.H., Majd, V.J.: A Robust H_∞ Non-PDC design scheme for singularly perturbed T-S fuzzy systems with immeasurable state variables. IEEE Trans. Fuzzy Syst. **23**(3), 525–541 (2015)
5. Ban, X., Gao, X.Z., Huang, X.: Graphical image analysis of Takagi-Sugeno control systems in the frequency domain. Fuzzy Sets Syst. **171**(1), 93–105 (2011)

6. Basil, M.A., Agustén, J., Ramón, G.L.: Fuzzy optimal control using generalized Takagi-Sugeno model for multivariable nonlinear systems. Appl. Soft Comput. **30**, 205–213 (2015)
7. Ben, Zina H., Allouche, M., Souissi, M., Chaabane, M., Chrifi-Alaoui, L.: Robust sensor fault-tolerant control of induction motor drive. Int. J. Fuzzy Syst. **19**(1), 155–166 (2017)
8. Blanke M., Kinnaert M., Lunze J., and Staroswiecki M. Diagnosis and Fault-Tolerant Control. Springer, Berlin (2003). ISBN 3-540-01056-4
9. Bouguila N., Jamel W., Khedher A., Ben Othman K.: Multiple observer design for a non-linear Takagi-Sugeno system submitted to unknown inputs and outputs IET Signal Process, pp. 1–11 (2013). https://doi.org/10.1049/iet-spr.2012.0398
10. Chang, W.J., Ku, C.C., Huang, P.H.: Robust fuzzy control for uncertain stochastic time-delay Takagi-Sugeno fuzzy models for achieving passivity. Fuzzy Sets Syst. **161**(15), 2012–2032 (2010)
11. Chen, J., Patton, R.J.: Fault-tolerant control systems design using the linear matrix inequality approach. In: 6th European Control Conference. Porto, Portugal (2001)
12. Dong, J., Yang, G.H., Zhang, H.: Stability analysis of T-S fuzzy control systems by using set theory. IEEE Trans. Fuzzy Syst. **23**(4), 827–841 (2015)
13. Elleuch, I., Khedher, A., Ben Othman, K.: State and faults estimation based on proportional integral sliding mode observer for uncertain Takagi-Sugeno fuzzy systems and its application to a turbo-reactor. Int. J. Fuzzy Syst. **19**(6), 1768–1781 (2017). https://doi.org/10.1007/s40815-017-0365-3
14. Emanoel, R.Q., Chaves, J., Andre, F.O., Dantas, A., Maitelli, A.L.: Unknown input observer-based actuator and sensor fault estimation technique for uncertain discrete time Takagi-Sugeno systems. Int. J. Control Autom. Syst. **19**(7), 2444–2454 (2021)
15. Haj, B.I., Chaabane, M., Driss, M.: Fault-tolerant control for T-S fuzzy descriptor systems with sensor faults: an LMI approach. Int. J. Fuzzy Syst. **19**(2), 516–527 (2017)
16. Hmidi, R., Ben Brahim, A., Dhahri, S., Ben Hmida, F., Sellami, A.: Sliding mode fault-tolerant control for Takagi-Sugeno fuzzy systems with local nonlinear models: application to inverted pendulum and cart system. Trans. Inst. Meas. Control **43**(4) (2021)
17. Hongyi, L., Shen, Y., Yingnan, P., Hak-Keung, L.: Model reduction for interval type-2 Takagi-Sugeno fuzzy systems. Automatica **61**, 308–314 (2015)
18. Jamel, W., Khedher, A., Ben Othman, K.: Observer design and active fault tolerant control for Takagi-Sugeno systems affected by sensors faults. Int. J. Model. Identification Control (IJMIC) 27(1), 22–30 (2017)
19. Jia, Q., Wu, L., Li, H.: Robust actuator fault reconstruction for Takagi-sugeno fuzzy systems with time-varying delays via a synthesized learning and luenberger observer. Int. J. Control Autom. Syst. **19**, 799–809 (2021). https://doi.org/10.1007/s12555-019-0747-4
20. Jian, H., Xiuhua, L., Xianwen, G., Xinjiang, W.: Intermediate observer-based robust distributed fault estimation for nonlinear multiagent systems with directed graphs. IEEE Trans. Indus. Inf. **16**(12), 7426–7436 (2020)
21. Jianglin, L., Ron, P.: Asymptotic estimation of state, fault and perturbation for nonlinear systems and its fault-tolerant control application. Int. J. Control, Autom. Syst. **19**(3), 1175–1182 (2021)
22. Johansen, T.A., Foss, A.B.: Nonlinear local model representation for adaptive systems. In: Singapore International Conference on Intelligent Control and Instrumentation. Singapore (1992)
23. Kalhor, A., Babak, N.A., Lucas, C.: Evolving Takagi-Sugeno fuzzy model based on switching to neighboring models. Appl. Soft Comput. **13**(2), 939–946 (2013)
24. Kamal, E., Aitouche, A., Ghorbani, R., Bayart, M.: Unknown input observer with fuzzy fault tolerant control for wind energy system. In: 8th IFAC Symposium on Fault Detection, Supervision and Safety of Technical Processes (SAFEPROCESS). Mexico City, Mexico (2012)
25. Kazumune, H., Michelle, S.Ch., Dimos, V.D.: Distributed l1-state-and-fault estimation for multiagent systems. IEEE Trans. Control Netw. Syst. **7**(2), 699–710 (2020)
26. Kharrat, D., Gassara, H., El Hajjaji, A., Chaabane, M.: Adaptive observer and fault tolerant control for takagi-sugeno descriptor nonlinear systems with sensor and actuator faults. Int. J. Control, Autom. Syst. **16**(3), 972–982 (2018). https://doi.org/10.1007/s12555-017-0546-8

27. Khedher, A., Ben, Othman K., Maquin, D., Benrejeb, M.: Sensor fault estimation for nonlinear systems described by Takagi-Sugeno models. Int. J. Trans. Syst. Signal Devices, Issues Syst. Signal Devices **6**(1), 1–18 (2011)
28. Khedher, A., Ben Othman, K., Maquin, D., Benrejeb, M.: Active sensor faults tolerant control for Takagi-Sugeno multiple models. In: 6th WSEAS International Conference on dynamical systems and control (CONTROL' 10) Sousse. Tunisia (2010)
29. Korbicz, J., Koscielny, J., Kowalczuk, Z., Cholewa, W.: Fault Diagnosis. Models, Artificial Intelligence, Applications. Springer, Berlin (2004) Intelligence, Applications. Springer, Berlin (2004)
30. Kukurowski, N., Pazera, M., Witczak, M., López-Estrada, F.R., Bravo Cruz, T.I.: Fault-tolerant tracking control for Takagi–Sugeno fuzzy systems under actuator and sensor faults. In: Bartoszewicz, A., Kabzinski, J., Kacprzyk, J. (eds.) Advanced, Contemporary Control. Advances in Intelligent Systems and Computing, vol 1196. Springer, Cham (2020). https://doi.org/10.1007/978-3-030-50936-1_119
31. Kumaresan, N., Ratnavelu, K.: Optimal control for stochastic linear quadratic singular neuro Takagi-Sugeno fuzzy system with singular cost using genetic programming. Appl. Soft Comput. **24**, 1136–1144 (2014)
32. Liang, Y., Liaw, D., Lee, T.: Reliable control of nonlinear systems. **45**, 706–710 (2000)
33. Liao, F., Wang, J., Yang, G.: Reliable robust flight tracking control: an lmi approach. IEEE Trans. Control Syst. Technic **10**, 76–89 (2002)
34. Makni, S., Bouattour, M., El Hajjaji, A., Chaabane, M.: Robust fault tolerant control based on adaptive observer for Takagi-Sugeno fuzzy systems with sensor and actuator faults: application to single-link manipulator. Trans. Inst. Meas. Control **42**(12) (2020)
35. Moodi, H., Farrokhi, M.: Robust observer-based controller design for Takagi-Sugeno systems with nonlinear consequent parts. Fuzzy Sets Syst. **273**, 141–154 (2015)
36. Navarbaf, A., Khosrowjerdi, M.J.: Fault-tolerant controller design with fault estimation capability for a class of nonlinear systems using generalized Takagi-Sugeno fuzzy model. Trans. Inst. Meas. Control **41**(15) (2019)
37. Orjuela, R., Marx, B., Ragot, J., Maquin, D.: On the simultaneous state and unknown inputs estimation of complex systems via a multiple model strategy. IET Control Theory Appl. **3**(7), 877–890 (2009)
38. Ouzaz, M., El Assoudi, A., Soulami, J., El Yaagoubi, E.H.: Simultaneous state and fault estimation for Takagi-Sugeno implicit models with Lipschitz constraints. Int. J. Optim. Control: Theories Appl. **11**(1), 100–108 (2021)
39. Qu, Z., Ihlefeld, C.M., Yufang, J., Saengdeejing, A.: Robust fault-tolerant self-recovering control of nonlinear uncertain systems. Automatica **39**, 1763–1771 (2003)
40. Radu-Emil, P., Horatiu-Ioan, F., Mircea-Bogdan, R., Emil, M.P., Preitl, S., Dragos, C.A.: Online identification of evolving Takagi-Sugeno-Kang fuzzy models for crane systems. Appl. Soft Comput. **24**, 1155–1163 (2014)
41. Sun, S., Wang, X., Zhang, H. Xie, X.: A new method of fault estimation and tolerant control for fuzzy systems against time-varying delay. Nonlinear Anal. Hybrid Syst. **38** (2020)
42. Takagi, M., Sugeno, M.: Fuzzy identification of systems and its application to modelling and control. IEEE Trans. Syst. Man Cybern. **15**(1), 116–132 (1985)
43. Tanaka, K., Ikeda, T., He, Y.Y.: Fuzzy regulators and fuzzy observers: relaxed stability conditions and LMI-based design. IEEE Trans. Fuzzy Syst. **6**(1), 250–256 (1998)
44. Wang, C., Dong, J., Yang, G., Kang, H.: Fuzzy fault tolerant control for nonlinear systems with sensor faults. In: Proceeding of the 32nd Chinese Control Conference. Xi'an China (2013)
45. Witczak, M., Pazera, M., Kukurowski, N., Bravo Cruz, T.I., Theilliol, D.: Robust actuator and sensor fault estimation for Takagi-Sugeno fuzzy systems under ellipsoidal bounding. In: IEEE International Conference on Fuzzy Systems (FUZZ-IEEE) (2020)
46. Xiuhua, L., Xianwen, G., Jian, H.: Distributed fault estimation for a class of nonlinear multiagent systems. IEEE Trans. Syst. Man Cybern. Syst. **50**(9), 3382–3390 (2020)
47. Zadeh, L.A.: Fuzzy sets information and control **8**, 338–353 (1965)

48. Zhang, D., Cai, W., Xie, L., Wang, Q.G.: Nonfragile distributed filtering for T-S fuzzy systems in sensor networks. IEEE Trans. Fuzzy Syst. **23**(5), 1883–1890 (2015)
49. Zhao, D., Hak Keung, L., Yueyang, L., Ding, S.X., Shuai, L.: A novel approach to state and unknown input estimation for Takagi–Sugeno fuzzy models with applications to fault detection. IEEE Trans. Circuits Syst. I: Regular Papers **67**(6), 2053–2063 (2020)

On Observer Design of Systems Based on Renewable Energy

Omar Naifar and Ghada Boukettaya

Abstract This chapter is devoted to the comparative study between two observation techniques namely: the adaptive observer and the interconnected observer applied to a wind energy conversion system (WECS) based induction machine (IM). It has been found that the introduced interconnected observer exhibits higher performance than the traditional adaptive one. Whereas, in the second part of this chapter, an adaptive interconnected observer is applied for both IM and PMSM based WECS. Such observer is robust and it compensates the effects of parametric variations.

Keywords Adaptive observer · Interconnected observer · Adaptive interconnected observer · WECS

1 Introduction

The use of wind energy (WE) has grown swiftly in North America, Europe and Asia. The Global Wind Energy Council (GWEC) mentioned that the total capacity of WE, established in 2012, surpassed 44 GW worldwide [1]. By 2020, the objective is to have 20% of its claimed electricity supplied by wind energy [2].

There are two basic techniques to synthesize observers which are used for sensorless control. The first technique is the free-model technique which we can cite as the heuristic technique like artificial intelligence [3–5] and the technique based on the machine geometry [6–8]. The second technique is based on the IM dynamic model. It uses automated tools to synthesize linear or nonlinear observers. In the literature, there are several kinds of such observers such as the extended Kalman filter [6, 9], the high gain interconnected observer [10–12] and the adaptive observers [13, 14].

O. Naifar (✉)
Control and Energy Management Laboratory, National School of Engineering of Sfax, University of Sfax, 3038 Sfax, Tunisia

G. Boukettaya
Reaserch Laboratory of Renewable Energy and Electric System (LSEER) National School of Engineers of Sfax (ENIS), 3038 Sfax, Tunisia

O. Naifar and A. Ben Makhlouf (eds.), *Advances in Observer Design and Observation for Nonlinear Systems*, Studies in Systems, Decision and Control 410,
https://doi.org/10.1007/978-3-030-92731-8_8

The work presented in this chapter was inspired by the method used in Refs. [15, 16] to compare an interconnected high-gain observer and an adaptive observer applied to IM. The speed observer based on the speed adaptation law was first proposed. Next, a speed observer based on the interconnected observer theory was described. Then, the two observers were compared, to test their robustness against parameter variations. Some simulation results were provided to demonstrate the effectiveness of the interconnected observer in sensorless control for the wind energy conversion system especially under inductive parameter variations.

Such observers are very sensitive to parametric variations especially the stator resistance. It is mentioned that the value of stator resistance is required for stator flux estimation. Its variation due to frequency or temperature affects the scheme performance. To overcome this problem, we have proposed in the fourth part of this chapter a stator resistance estimator. This estimator is included to compensate for the effects of the stator resistance variations. A stability-analysis method of the suggested observer was introduced and discussed.

In the fifth part of this chapter paper, an online estimator is included to compensate the effects of parametric variations for PMSM. The design of the proposed observer provides an extended model of the machine so that PMSM parameters behave as state variables, since the mechanical speed is not a state variable in the electric state model, it is considered as a parameter. For the proof of the observer, the stability-analysis method in the sense of Lyapunov was introduced and discussed. Simulation results were presented in order to validate and show the performance of the proposed observer.

The proposed observer used for both IM and PMSM is named as an adaptive interconnected observer because of the speed adaptation law and the interconnection form between each observer. The advantages of this approach are defined as follows:

- Considering a state variable as an adjustable parameter in the state matrix and so the possibility to use linear observations techniques on nonlinear systems.
- The possibility to extend the nonlinear system in a form so that parameters act as state variables with slight variation. This approach allows parametric estimation of the system.

2 Modeling

2.1 Induction Machine Model in the Stationary Reference Frame (α, β)

The electrical state model of the induction motor in the stationary reference frame (α, β) is presented by the system (1) [17]:

$$\begin{cases} \dot{x}_{\alpha\beta} = f\left(x_{\alpha\beta}\right) + Bu_{\alpha\beta} \\ y_{\alpha\beta} = Cx_{\alpha\beta} \end{cases} \tag{1}$$

where: $x_{\alpha\beta} = [i_{s\alpha} i_{s\beta} \phi_{r\alpha} \phi_{r\beta}]^T$ is the state vector, $y_{\alpha\beta} = \begin{bmatrix} i_{s\alpha} \\ i_{s\beta} \end{bmatrix}$ is the output vector and $u_{\alpha\beta} = \begin{bmatrix} u_{s\alpha} \\ u_{s\beta} \end{bmatrix}$ is the control vector.

$$f(x_{\alpha\beta}) = \begin{pmatrix} -\gamma_1 i_{s\alpha} + \frac{\gamma_2}{\tau_r}\phi_{r\alpha} + \gamma_2 p\Omega\phi_{r\beta} \\ -\gamma_1 i_{s\beta} - \gamma_2 p\Omega\phi_{r\alpha} + \frac{\gamma_2}{\tau_r}\phi_{r\beta} \\ \frac{L_m}{\tau_r} i_{s\alpha} - \frac{1}{\tau_r}\phi_{r\alpha} - p\Omega\phi_{r\beta} \\ \frac{L_m}{\tau_r} i_{s\beta} - \frac{1}{\tau_r}\phi_{r\beta} + p\Omega\phi_{r\alpha} \end{pmatrix}$$ is the state matrix, $B = \begin{pmatrix} m_1 & 0 \\ 0 & m_1 \\ 0 & 0 \\ 0 & 0 \end{pmatrix}$

is the input matrix, and $C = \begin{pmatrix} 1\,0\,0\,0 \\ 0\,1\,0\,0 \end{pmatrix}$ is the output matrix. The parameters σ, τ_r, γ_1 and γ_2 are defined by:

$$\sigma = 1 - \left(L_m^2/L_s L_r\right),\ \tau_r = \frac{L_r}{R_r},\ \gamma_1 = \frac{\frac{L_m^2}{L_r^2}R_r + R_s}{\sigma L_s},\ m_1 = \frac{1}{\sigma L_s} \text{ and } \gamma 2 = L_m/L_r L_s \sigma.$$

L_s: stator inductance, (R_r, R_s):rotor and stator resistances, σ: leakage coefficient, τ_r:rotor time constant,.L_r.: rotor inductance, L_m: mutual inductance, p: number of pole pairs, J: rotor moment of inertia and Ω is the speed of the machine.

The IM physical operation domain D_1 in the reference frame (α, β) is defined as follows:

$$D_1 = \{\ X \in \mathbb{R}^5 \| i_{s\alpha}| \le i_{s\alpha}^{max}, |i_{s\beta}| \le i_{s\beta}^{max}, |\Omega| \le \Omega^{max}, |\phi_{r\alpha}| \le \phi_{r\alpha}^{max}, |\phi_{r\beta}| \le \phi_{r\beta}^{max}\} \tag{2}$$

with $X_{\alpha\beta} = \left[i_{s\alpha} i_{s\beta} \Omega \phi_{r\alpha} \phi_{r\beta}\right]^T$ and $i_{s\alpha}^{max}$,$i_{s\beta}^{max}$,Ω^{max},$\phi_{r\alpha}^{max}$ and $\phi_{r\beta}^{max}$ are the actual maximum values for currents, speed and flux.

2.2 *Induction Machine Model in the Park Frame*

According to [18], the rotation matrix N defined in (3) allows expressing model (1) in the Park frame, we have:

$$\begin{bmatrix} x_d \\ x_q \end{bmatrix} = N(\rho)^T \begin{bmatrix} x_\alpha \\ x_\beta \end{bmatrix} \tag{3}$$

with $N(\rho) = \begin{pmatrix} \cos(\rho) & -\sin(\rho) \\ \sin(\rho) & \cos(\rho) \end{pmatrix}$ and $\rho = \begin{pmatrix} \frac{\phi_{r\alpha}}{\phi_{r\beta}} \end{pmatrix}$ is the angular position of the rotating.

Then, the induction machine state model in the Park frame is represented by:

$$\begin{cases} \frac{d}{dt}x_{dq} = f(x_{dq}) + Bu_{dq} \\ y_{dq} = Cx_{dq} \end{cases} \tag{4}$$

where: $x_{dq} = [i_{sd} i_{sq} \phi_{rd} \phi_{rq}]^T$ is the state vector, $f(x_{dq}) = \begin{bmatrix} -\gamma_1 i_{sd} + \frac{\gamma_2}{\tau_r}\phi_{rd} + \omega_s i_{sq} + p\Omega\gamma_2\phi_{rq} \\ -\gamma_1 i_{sq} + \frac{\gamma_2}{\tau_r}\phi_{rq} - \omega_s i_{sd} - p\Omega\gamma_2\phi_{rd} \\ \frac{L_m}{\tau_r} i_{sd} - \frac{1}{\tau_r}\phi_{rd} - (p\Omega - \omega_s)\phi_{rq} \\ \frac{L_m}{\tau_r} i_{sq} - \frac{1}{\tau_r}\phi_{rq} + (p\Omega - \omega_s)\phi_{rd} \end{bmatrix}$, $y_{dq} = \begin{bmatrix} i_{sd} \\ i_{sq} \end{bmatrix}$ is the output vector and $u_{dq} = \begin{bmatrix} u_{sd} \\ u_{sq} \end{bmatrix}$ is the control vector.

The IM physical operation domain D_2 is defined as follows:

$$D_2 = \left\{ X_1 \in \mathbb{R}^5 \| i_{sd} | < i_{sd} \max, |i_{sq}| < i_{sq}^{\max} |\Omega| < \Omega^{\max}, |\phi_{rd}| < \phi_{rd}^{\max}, |\phi_{rd}| < \phi_{rd}^{\max} \right\} \tag{5}$$

with $X_{dq} = \left[i_{sd} i_{sq} \Omega \phi_{rd} \phi_{rq} \right]^T$ and i_{sd}^{max},i_{sq}^{max},Ω^{max},ϕ_{rd}^{max} and $\phi_{rd}^{\max}$ are the actual maximum values for currents, speed and flux.

We can assume a perfect orientation of the frame dq, the axis d coincides with the rotor flux vector, therefore, the quadrature component of flux as well as its derivative will be cancelled ($\phi_{rq} = 0 \Rightarrow \dot{\phi}_{rq} = 0$). The nonlinear model of the induction machine in the Park frame with flux orientation is presented as follow:

$$\begin{cases} \frac{d}{dt}x'_{dq} = f(x'_{dq}) + B' u_{dq} \\ y'_{dq} = C' x'_{dq} \end{cases} \tag{6}$$

where: $x'_{dq} = [i_{sd} i_{sq} \phi_{rd}]^T$ is the state vector, $f(x'_{dq}) = \begin{bmatrix} -\gamma_1 i_{sd} + \frac{\gamma_2}{\tau_r}\phi_{rd} + \omega_s i_{sq} \\ -\gamma_1 i_{sq} - \omega_s i_{sd} - p\Omega\gamma_2\phi_{rd} \\ \frac{L_m}{\tau_r} i_{sq} + (p\Omega - \omega_s)\phi_{rd} \end{bmatrix}$, $y'_{dq} = \begin{bmatrix} i_{sd} \\ i_{sq} \end{bmatrix}$ is the output vector and $u_{dq} = \begin{bmatrix} u_{sd} \\ u_{sq} \end{bmatrix}$

is the control vector. $B^{'} = \begin{pmatrix} m_1\,0 \\ 0\,m_1 \\ 0\,0 \end{pmatrix}$ is the input matrix, and $C^{'} = \begin{pmatrix} 1\,0\,0 \\ 0\,1\,0 \end{pmatrix}$ is the output matrix.

2.3 *Permanent Magnet Synchronous Machine Model*

The mathematical model of the PMSM in a synchronous rotating frame is described by [19]:

$$\begin{cases} \dfrac{di_d}{dt} = -R_s L_s^{-1} i_d + p\Omega i_q + L_s^{-1} v_d \\ \dfrac{di_q}{dt} = -R_s L_s^{-1} i_q - p\Omega i_d - p\phi_f L_s^{-1}\Omega + L_s^{-1} v_q \\ \dfrac{d\Omega}{dt} = \dfrac{1}{J}(T_w - p\phi_f i_q - f_v\Omega) \end{cases} \tag{7}$$

with: R_s is the stator resistance, L_s is the stator inductance, ϕ_f is the permanent magnet flux linkage, i_d, i_q is the stator currents, Ω is the rotor mechanical speed, p is the number of pare pole, f_v is the viscous friction coefficient, J is the moment of inertia and T_w is the wind torque.

The PMSM physical operation domain D is defined as follows:

$$D_3 = \left\{ X_2 \in \mathbb{R}^5 \| i_d | < i_d^{\max}, | i_q | < i_q^{\max} | \Omega | < \Omega^{\max}, | R_s | < R_s^{\max}, | L_s^{-1} | < L_s^{-1} \underset{3}{\max} \right\} \tag{8}$$

with $X = \left[i_d i_q \Omega R_s L_s^{-1} \right]^T$ and $i_d^{max}, i_q^{max}, \Omega^{max}, R_s^{max}$ and $L_s^{-1\max}$ are the actual maximum values for currents, speed stator resistance and stator inductance.

2.4 *Wind Turbine Modeling*

We adopt a horizontal axis wind turbine with three blades in length R and generally driving a generator through a G gain speed multiplier.

The mechanical rotation speed is determined from the fundamental relation of the dynamics as:

$$J\frac{d\Omega}{dt} = T_w - T_{em} - f\Omega \tag{9}$$

where Ω is the machine speed, T_{em} is the electromagnetic torque, T_w is the wind torque and f is the friction coefficient.

In the case of induction machine, T_{em} is defined by:

$$T_{em} = p\frac{L_m}{L_r}\left(i_{sq}\phi_{rd} - i_{sd}\phi_{rq}\right) \tag{10}$$

In the case permanent magnet synchronous machine, T_{em} is defined by:

$$T_{em} = p\phi_f i_q \tag{11}$$

The mechanical torque of the wind turbine T_w may be defined [20, 21]:

$$T_w = \frac{1}{2}\rho\pi\frac{R^3 v^2}{\lambda}C_p(\lambda, \beta) \tag{12}$$

The power coefficient C_p is the aerodynamic efficiency of the wind turbine. It depends on the characteristic of the turbine [22]: the speed ratio λ and the angle of orientation of the blades β are defined as follows:

$$C_p(\lambda, \beta) = 0.53\left[\frac{151}{\lambda_i} - 0.58\beta - 0.002\beta^{2.14} - 13.2\right]exp\left(\frac{-18.4}{\lambda_i}\right) \tag{13}$$

with

$$\lambda_i = \left(\frac{1}{\lambda - 0.02\beta} - \frac{0.003}{\beta^3 + 1}\right)^{-1} \tag{14}$$

$$\lambda = \frac{R\Omega_t}{v} \tag{15}$$

3 A Comparative Study Between a High Gain Interconnected Observer and An Adaptive Observer

In this section, a comparison study is carried out between two observation approaches dedicated to speed control strategies of induction machine (IM) under parametric variations, such that: (i) the adaptive observer approach which is based on the speed adaptation law and (ii) the interconnected observer that offers robustness and stability of the system with reduced CPU time. The comparison study is achieved considering four performance criteria: the stability, the robustness to the variations of the machine inductances, the robustness to the variations of the machine resistances, and the feasibility of the torque estimation. It has been found that the introduced interconnected observer exhibits higher performance than the traditional adaptive one, to the above-cited comparison criteria.

3.1 Adaptive Observer

The goal of the adaptive observer is to jointly estimate the state and the unknown parameters of the parametric system model. Parametric adaptation law is used by the adaptive observer which is derived from a Lyapunov function integrating the output error. This technique can be applied to classes of linear or non-linear systems. The adaptive observer is widely used to estimate the mechanical speed for the control of electrical machines. For the design of the adaptive observer, we use the induction machine model as it is presented in Eq. (1).

In the referential (α, β), the adaptive observer is presented by Eq. (16) [16]:

$$\begin{cases} \dot{\hat{x}} = A(\hat{\Omega})\hat{x} + L(\hat{y} - y) + Bu \\ \hat{y} = C\hat{x} \end{cases} \tag{16}$$

where $\hat{x}$,$\hat{y}$ and $\hat{\Omega}$ are respectively the estimation of x, y and Ω, $L = \begin{pmatrix} l_{11} & l_{12} \\ l_{21} & l_{22} \\ l_{31} & l_{32} \\ l_{41} & l_{42} \end{pmatrix}$ is the gain matrix.

Remark 1 The proposed observer in this section is the Full order Luenberger observer, but it is named as an adaptive observer because of the speed adaptation law. The advantage of this approach (considering the speed as a parameter in the state matrix) allows using linear observations techniques on nonlinear systems.

The estimation error vector e is defined by:

$$e = \hat{x} - x = [e_{is\alpha} e_{is\beta} e_{\phi r\alpha} e_{\phi r\beta}]^T \tag{17}$$

The dynamics of the estimation error under parametric variation is given by this system:

$$\begin{cases} \dot{e} = \left(A\left(\hat{\Omega}\right) + LC\right)e + (\Delta A(\Omega) + \delta A(\Omega))\hat{x} \\ \varepsilon = Ce \end{cases} \tag{18}$$

where: ε is the output term, $\Delta A(\Omega) = A\left(\hat{\Omega}\right) - A(\Omega)$ and $\delta A(\Omega)$ is the uncertain term of $A(\Omega)$.To ensure the observer convergence, it is required that system (18) is stable. The Lyapunov function is chosen in the following form:

$$V_a = \Upsilon + \lambda(\Delta\Omega)^2 \tag{19}$$

where $\Upsilon = e^T Pe$, $\Delta\Omega = \hat{\Omega} - \Omega$, P is a matrix such as $P^T = P > 0$, and λ is a positive weighting coefficient.

To ensure the convergence of the observer, it is required that the derivative of V_a must be negative and definite. Thus we have:

$$\dot{V}_a = \sigma_1 + \sigma_2 < 0 \tag{20}$$

Where: $\sigma_1 = e^T\Big\{(A(\hat{\Omega}) + LC)^T P + P(A(\hat{\Omega}) + LC)\Big\}e + 2e^T P\delta A(\Omega)\hat{x}$

and $\sigma_2 = 2e^T P\Delta A(\Omega)\hat{x} + 2\lambda\Delta\Omega\frac{d}{dt}\hat{\Omega}$.

The observer (16) is stable if it satisfies conditions (21) and (22):

$$\sigma_1 < 0 \tag{21}$$

and

$$\sigma_2 = 0 \tag{22}$$

To prove of condition (19), it is required that $(A\left(\hat{\Omega}\right) + LC)$ is stable, then $\forall Q = Q^T > 0$, $\exists P = P^T > 0$ such as $\left(A\left(\hat{\Omega}\right) + LC\right)^T P + P\left(A\left(\hat{\Omega}\right) + LC\right) = -Q$. According to [16], the parameters of the L matrix gains are given as follows:

$$l_{11} = l_{22} < 0, l_{12} = l_{21} = 0 \text{ and } \left\{\begin{array}{l} l_{31} = l_{42} = -\frac{L_m}{\tau_r} \\ l_{32} = l_{41} = 0 \end{array}\right. \tag{23}$$

Assumption (24) is justified by the fact that $\hat{\Omega}$ is a regularly persistent input for system (16) and the machine state and parameters are bounded:

$$\left\{\begin{array}{l} \|P\| \leq \xi_1 \\ \left\|\hat{x}\right\| \leq \xi_2 \\ \|\delta A(\Omega)\| \leq \zeta_3 \\ \lambda_{min}(P)\|e\|^2 \leq \Upsilon \leq \lambda_{max}(P)\|e\|^2 \end{array}\right. \tag{24}$$

where ξ_1,ξ_2,ζ_3,$\lambda_{min}(P)$ and $\lambda_{max}(P)$ are real positive numbers with $\lambda_{min}(.)$, $\lambda_{min}(.)$ present respectively the minimum and the maximum eigenvalues.

Equation (21) becames:

$$\begin{aligned} & -\eta_1\Upsilon + \eta_2\sqrt{\Upsilon} \\ & \leq -\eta_1\Upsilon + \eta_2\frac{\Upsilon}{\sqrt{\Upsilon}} \\ & \leq -\eta_1\Upsilon + \eta_2\Upsilon\frac{1}{\sqrt{\lambda_{\min}(P)e}} \\ & \leq -\eta_1\Upsilon\left(1 - \frac{\eta_2}{\eta_1\sqrt{\lambda_{\min}(P)e}}\right) \end{aligned} \tag{25}$$

where $\eta_1 = \frac{\lambda_{min}(Q)}{\lambda_{max}(P)}$, $\eta_2 = \frac{2\xi_1\xi_2\xi_3}{\sqrt{\lambda_{min}(P)}}$.

If the machine parameters are known, in this case $\eta_2 = 0$.In order to check the condition (21), we just choose $\eta_1 > 0$.If the machine parameters are variable, then in this case $\eta_2 \neq 0$. For σ_1 defined negative, it is required that the error estimate is always greater than $\frac{\eta_2}{\varepsilon\eta_1\sqrt{\lambda_{min}(P)}}$ with $\varepsilon \in]0, 1[$.

Then, Eq. (23) is developed as:

$$\sigma_1 < -(1-\varepsilon)\eta_1\Upsilon \tag{26}$$

According to [23], the speed adaptation law is developed from (22) and it is presented by:

$$\tfrac{d}{dt}\hat{\Omega} = \varsigma_1\tfrac{1}{2\lambda}p\left(e_{is\beta}\hat{\Phi}_{r\alpha} - e_{is\alpha}\hat{\Phi}_{r\beta}\right) - \varsigma_2\frac{1}{2\lambda}p\left(e_{\phi r\beta}\hat{\Phi}_{r\alpha} - e_{\phi r\alpha}\hat{\Phi}_{r\beta}\right) \tag{27}$$

where ς_1 and ς_2 are two positive constants. Knowing that the established observer output is $Ce = \left[e_{is\alpha}, e_{is\beta}\right]$, the second term of Eq. (27) is unknown and thus considered as being a disturbance. To cancel the effect of this disturbance, a Proportional Integral regulator is used, as shown in Fig. 1.

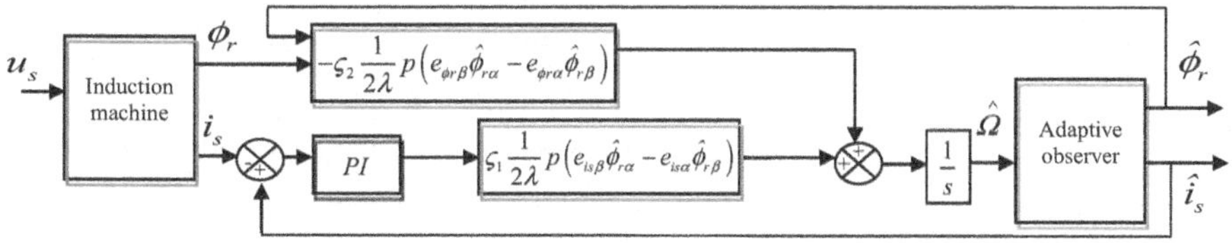

Fig. 1 Block diagram of the designed adaptive observer

3.2 Interconnected Observer

It is well known that there are no systematic methods for the design of observers for a given non-linear system. However, several observation methods which are available depending on specific characteristics of the system have been studied. In particular, the nonlinear system considered can be seen as an interconnection between several subsystems, where each subsystem satisfying certain conditions, an observer can be synthesized. The idea of the interconnected observer is to design an observer for the whole nonlinear system from the synthesis of separate observers for each subsystem with the hypothesis that the statements of other subsystems are available for each observer. Certain assumptions are then used to prove the convergence of all observers.

3.2.1 Observer Design

The induction machine model (1) can be decomposed into two affine-state interconnected subsystems as follows:

$$\begin{cases} \dot{x}_1 = A_1(x_2)x_1 + g(u, x_2, x_1) \\ y_1 = C_1 x_1 \end{cases} \tag{28}$$

$$\begin{cases} \dot{x}_2 = A_2(x_1)x_2 + \varphi(u, y) \\ y_2 = C_2 x_2 \end{cases} \tag{29}$$

where: $x_1 = [i_{s\alpha}\Omega]^T$ is the state vector of the first subsystem; $x_2 = [i_{s\beta}\phi_{r\alpha}\phi_{r\beta}]^T$ is the state vector of the second subsystem; $A_1(x_2) = \begin{pmatrix} 0 & \gamma_2 p\phi_{r\beta} \\ 0 & 0 \end{pmatrix}$ is the state matrix of the first subsystem; $A_2(x_1) = \begin{pmatrix} -\gamma_1 & -\gamma_2 p\Omega & \frac{\gamma_2}{\tau_r} \\ 0 & \frac{-1}{\tau_r} & -p\Omega \\ 0 & p\Omega & \frac{-1}{\tau_r} \end{pmatrix}$ is the state matrix of the second subsystem; $C_1 = \begin{pmatrix} 1 & 0 \end{pmatrix}$ and $C_2 = \begin{pmatrix} 1 & 0 & 0 \end{pmatrix}$ are the vector outputs; $g(u, x_2, x_1) = \begin{pmatrix} -\gamma_1 i_{s\alpha} + \frac{\gamma_2}{\tau_r}\phi_{r\alpha} + m_1 u_{s\alpha} \\ -m\left(\phi_{r\alpha} i_{s\beta} - \phi_{r\beta} i_{s\alpha}\right) - c\Omega + \frac{1}{J}T_w \end{pmatrix}$ and $\varphi(u, y) = \begin{pmatrix} m_1 u_{s\beta} \\ \frac{L_m}{\tau_r} i_{s\alpha} \\ \frac{L_m}{\tau_r} i_{s\beta} \end{pmatrix}$ is an input–output injection term.

The interconnected observer is based on the interconnection between several observers; it requires some properties, especially the property of input persistence (See [10] for more information).

Assumption 1 x_2 and x_1 are respectively inputs for subsystems (28) and (29).

Assumption 2

1. $A_1(x_2)$ is globally lipschitz with respect to x_2
2. $g(u, x_2, x_1)$ is globally lipschitz with respect to x_2 uniformly with respect to the pair (u, x_1).
3. $A_2(x_1)$ is globally lipschitz with respect to x_1

A high gain observer for the system of Eqs. (28) and (29) given respectively by system (30) and (31):

$$\begin{cases} \dot{\hat{x}}_1 = A_1(\hat{x}_2)\hat{x}_1 + g(u, \hat{x}_2, \hat{x}_1) + M_1(\hat{x}_2, \theta_1)(y_1 - \hat{y}_1) \\ \hat{y}_1 = C_1\hat{x}_1 \end{cases} \tag{30}$$

$$\begin{cases} \dot{\hat{x}}_2 = A_2(\hat{x}_1)\hat{x}_2 + \varphi(u, y) + M_2(\theta_2, \hat{x}_1)(y_2 - \hat{y}_2) \\ \hat{y}_2 = C_2\hat{x}_2 \end{cases} \tag{31}$$

where $\hat{x}_1 = [\hat{i}_{s\alpha}\hat{\Omega}]^T$ is the estimated vector of x_1 and $\hat{x}_2 = [\hat{i}_{s\beta}\hat{\phi}_{r\alpha}\hat{\phi}_{r\beta}]^T$ is the estimated vector of x_2; $A_1(\hat{x}_2) = \begin{pmatrix} 0 & \gamma_2 p\hat{\phi}_{r\beta} \\ 0 & 0 \end{pmatrix}$ is the estimated matrix of $A_1(x_2)$; $A_2(\hat{x}_1) = \begin{pmatrix} -\gamma_1 & -\gamma_2 p\hat{\Omega} & \frac{\gamma_2}{\tau_r} \\ 0 & \frac{-1}{\tau_r} & -p\hat{\Omega} \\ 0 & p\hat{\Omega} & \frac{-1}{\tau_r} \end{pmatrix}$ is the estimated matrix of $A_2(x_1)$ and $g(u, \hat{x}_2, \hat{x}_1)$ is the estimation term of $g(u, x_2, x_1)$.

The designed observer gain $M_1(\theta_1, \hat{x}_2)$ of system (28) is chosen as $M_1(\theta_1, \hat{x}_2) = \Gamma^{-1}(\hat{x}_2)S_1^{-1}C_1^T$.

Where: $\Gamma(\hat{x}_2) = \begin{pmatrix} 1 & 0 \\ 0 & \gamma_2 p\hat{\phi}_{r\beta} \end{pmatrix}$ and S_1 is the solution of the differential equation defined by [10]:

$$\dot{S}_1(\theta_1) = -\theta_1 S_1(\theta_1) - A_0^T S_1(\theta_1) - S_1(\theta_1)A_0 + C_1^T C_1 \tag{32}$$

with θ_1 is a positive constant and $A_0 = \begin{pmatrix} 0 & 1 \\ 0 & 0 \end{pmatrix}$.

$M_2(\theta_2, \hat{x}_1) = S_2^{-1}C_2^T$ is the second observer gain obtained by solving the following equation [10]:

$$\dot{S}_2(\theta_2, \hat{x}_1) = -\theta_2 S_2(\theta_2, \hat{x}_1) - A_2^T S_2(\theta_2, \hat{x}_1) - S_2(\theta_2, \hat{x}_1)A_2 + C_2^T C_2 \tag{33}$$

with θ_2 is a positive constant.

Figure 2 represents the architecture of the adopted interconnected observer applied to an induction machine.

To prove the convergence of the estimation error of the interconnected observers, we present in the following the stability analysis based on Lyapunov theory.

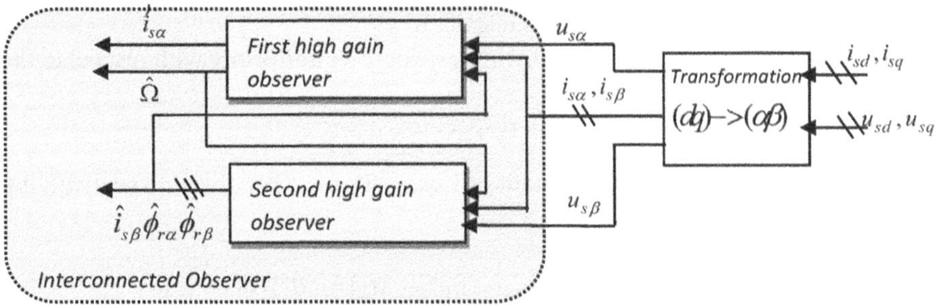

Fig. 2 Architecture of the IM interconnected observer

3.2.2 Stability Study of the Interconnected Observer Under Parametric Variation

Consider the estimation errors defined by:

$$e_1 = x_1 - \hat{x}_1 \text{and} e_2 = x_2 - \hat{x}_2 \tag{34}$$

The dynamics error e_1 is defined by:

$$\begin{aligned}\dot{e}_1 = [A_1(\hat{x}_2) - \Gamma^{-1}(\hat{x}_2)S_1^{-1}C_1^T C_1]e_1 + g(u, x_2, x_1) - g(u, \hat{x}_2, \hat{x}_1)\\ + [A_1(x_2) - A_1(\hat{x}_2)]x_1\end{aligned} \tag{35}$$

The dynamics error e_2 is defined by:

$$\dot{e}_2 = A_2(x_1)x_2 - A_2(\hat{x}_1)\hat{x}_2 - S_2^{-1}(\theta_2, \hat{x}_1)C_2^T C_2 e_2 \tag{36}$$

Now consider the Eqs. (35) and (36) with uncertainties on the IM parameters:

$$\begin{aligned}\dot{e}_1 = [A_1(\hat{x}_2) - \Gamma^{-1}(\hat{x}_2)S_1^{-1}C_1^T C_1]e_1 + g(u, x_2, x_1) - g(u, \hat{x}_2, \hat{x}_1) + \delta g(u, x_2, x_1)\\ +[A_1(x_2) - A_1(\hat{x}_2) + \delta A_1(x_2)]x_1\end{aligned} \tag{37}$$

and

$$\dot{e}_2 = [A_2(x_1) + \delta A_2(x_1)]x_2 + \delta\varphi(u, y) - A_2(\hat{x}_1)\hat{x}_2 - S_2^{-1}(\theta_2, \hat{x}_1)C_2^T C_2 e_2 \tag{38}$$

where: $\delta g(u, x_2, x_1)$,$\delta A_1(x_2)$, $\delta\varphi(u, y)$ and $\delta A_2(x_1)$ are respectively the uncertain terms of $g(u, x_2, x_1)$,$A_1(x_2)$, $\varphi(u, y)$ and $A_2(x_1)$.

Remark 3 According to Lemma 1, it is clear that $w = \hat{x}_2$ and $S(t) = S_1$ for subsystem (30), and for subsystem (31)$w = \hat{x}_1$ and $S(t) = S_2$.

Theorem 1 *Consider the interconnected IM model presented by Eqs. (28) and (29), system (30)–(31) is a high gain interconnected observer for system (28)–(29).*

Proof Considering the following coordinate change $\varepsilon_1 = \Gamma(\hat{x}_2)e_1$.

To check convergence, consider the following Lyapunov equation:

$$V_i = V_1 + V_2 \tag{39}$$

where $V_1 = \varepsilon_1^T S_1 \varepsilon_1$ and $V_2 = e_2^T S_2 e_2$.

By calculating the derivative of V_i along the trajectories ε_1 and e_2, $\dot{V}_i$ is presented as:

$$\begin{array}{c} \dot{V}_i = -\varepsilon_1^T\left[C_1^T C_1 + \theta_1 S_1\right]\varepsilon_1 - e_2^T\left[C_2^T C_2 + \theta_2 S_2\right]e_2 \\ +2\varepsilon_1^T S_1 \Gamma(\hat{x}_2)[g(u, x_2, x_1) - g(u, \hat{x}_2, \hat{x}_1) + \delta g(u, x_2, x_1)] \\ +2\varepsilon_1^T S_1 \Gamma(\hat{x}_2)\left[A_1(x_2) - A_1(\hat{x}_2) + \delta A_1(x_2)\right]x_1 + 2\varepsilon_1^T S_1 \dot{\Gamma}(\hat{x}_2)\Gamma^{-1}(\hat{x}_2)\varepsilon_1 \\ +2e_2^T S_2\left[A_2(x_1) - A_2(\hat{x}_1) + \delta A_2(x_1)\right]x_2 + 2e_2^T S_2 \delta\varphi(u, y) \end{array} \tag{40}$$

According to Lemma 1 and taking into account the initial conditions of the IM drive and the observer in the physical operation domain D_1, the following inequalities hold:

$$\begin{cases} \|S_1\| < k_1 \\ \left\|A_1(x_2) - A_1(\hat{x}_2)\right\| < k_2\|e_2\| \\ \|x_1\| < k_3 \\ \|x_2\| < k_4 \\ \left\|A_2(x_1) - A_2(\hat{x}_1)\right\| < k_5\|e_1\| \\ \left\|g_1(u, x_2, x_1) - g_1(u, \hat{x}_2, \hat{x}_1)\right\| < k_6\|e_2\| \\ \|S_2\| < k_7 \\ \left\|\Gamma(\hat{x}_2)\right\| < \rho_1 \\ \left\|\Gamma(\hat{x}_2)^{-1}\right\| < \rho_2 \\ \left\|\dot{\Gamma}(\hat{x}_2)\Gamma^{-1}(\hat{x}_2)\right\| < \rho_3 \\ \|\delta A_1(x_2)\| \leq \alpha_2 \\ \|\delta A_2(x_1)\| \leq \alpha_3 \\ \|\delta\varphi(u, y)\| \leq \alpha_4 \end{cases} \tag{41}$$

where $k_i, i \in \{1, \ldots\ldots, 7\}$ and $\rho_j, j \in \{1, 2, 3\}$ are positive constants. The computation of $k_i, i \in \{1, \ldots\ldots, 7\}$ are detailed in Annex.

Including the standards, Eq. (40) can be written as follows:

$$\dot{V}_i \leq -\theta_1\varepsilon_1^T S_1\varepsilon_1 - \theta_2 e_2^T S_2 e_2 + 2k_1k_6\rho_1\|\varepsilon_1\|\|e_2\| + 2k_1k_3\rho_1k_2\|e_2\|\|\varepsilon_1\|$$

$$+ 2k_1\rho_3\|\varepsilon_1\|\|\varepsilon_1\| + 2k_7k_4k_5\rho_2\|e_2\|\|\varepsilon_1\| + 2k_1\rho_1\alpha_1\|\varepsilon_1\|$$
$$+ 2k_3k_1\rho_1\alpha_2\|\varepsilon_1\| + 2(k_7\alpha_3k_4 + k_7\alpha_4)\|e_2\| \tag{42}$$

Since $\hat{x}_2$ and $\hat{x}_1$ are the inputs for subsystems (28) and (29) respectively. So from Lemma 1 $\exists$ real numbers $\lambda_{min}(S_1) > 0$,$\lambda_{\max}(S_1) > 0$,$\lambda_{min}(S_2) > 0$,$\lambda_{\max}(S_2) > 0$ such that:

$$\begin{cases} \lambda_{min}(S_1)\|\varepsilon_1\|^2 \le V_1 \le \lambda_{max}(S_1)\|\varepsilon_1\|^2 \\ \lambda_{min}(S_2)\|e_2\|^2 \le V_2 \le \lambda_{max}(S_2)\|e_2\|^2 \end{cases} \tag{43}$$

Using the inequalities (43) and by the fact that $\varepsilon_1 = \Gamma(\hat{x}_2)e_1$,it can be deduced that:

$$\begin{cases} \|e_1\| \le \rho_2 \dfrac{1}{\sqrt{\lambda_{min}(S_1)}}\sqrt{V_1} \\ \|e_2\| \le \dfrac{1}{\sqrt{\lambda_{min}(S_2)}}\sqrt{V_2} \\ \|\varepsilon_1\| \le \dfrac{1}{\sqrt{\lambda_{min}(S_1)}}\sqrt{V_1} \end{cases} \tag{44}$$

The derivative of V_i satisfy the following condition:

$$\begin{aligned} \dot{V}_i &\le -(\theta_1 - \tfrac{2k_1\rho_3}{\lambda_{min}(S_1)})V_1 - \theta_2 V_2 + \left(\tfrac{2k_1k_6\rho_1 + 2k_1k_3\rho_1k_2 + 2k_7k_4k_5\rho_2}{\sqrt{\lambda_{min}(S_1)\lambda_{min}(S_2)}}\right)\sqrt{V_1}\sqrt{V_2} \\ &+(2k_1\rho_1\alpha_1 + 2k_3k_1\rho_1\alpha_2)\frac{\sqrt{V_1}}{\sqrt{\lambda_{min}(S_1)}} + 2k_7\alpha_3k_4\frac{\sqrt{V_2}}{\sqrt{\lambda_{min}(S_2)}} \end{aligned} \tag{45}$$

On the other hand, Eq. (46) is checked:

$$\sqrt{V_1}\sqrt{V_2} \le \frac{1}{2}(V_1 + V_2) \tag{46}$$

By substituting (46) into (45),the result will be:

$$\dot{V}_i \le -\lambda_1 V_i + \lambda_2\psi\sqrt{V_i} \tag{47}$$

where $\lambda_1 = \min(\theta_1 - \frac{\lambda}{2} - \frac{2k_1\rho_3}{\lambda_{min}(S_1)}, \theta_2 - \frac{\lambda}{2})$ with $\lambda = \frac{2k_1k_6\rho_1 + 2k_1k_3\rho_1k_2 + 2k_7k_4k_5\rho_2}{\sqrt{\lambda_{min}(S_1)\lambda_{min}(S_2)}}$,$\lambda_2 = \max(\frac{2k_1\rho_1\alpha_1 + 2k_3k_1\rho_1\alpha_2}{\sqrt{\lambda_{min}(S_1)}}, \frac{2(k_7\alpha_3k_4 + k_7\alpha_4)}{\sqrt{\lambda_{min}(S_2)}})$ and $\psi > 0$ such that $\sqrt{V_1} + \sqrt{V_2} < \psi\sqrt{V_1 + V_2}$. Since $V_i = V_1 + V_2$ and taking into account (44) we can write:

$$\rho_2\lambda_{\min}(S_1)\|e_1\|^2 + \lambda_{\min}(S_2)\|e_2\|^2 \le V_i \le \rho_2\lambda_{\max}(S_1)\|e_1\|^2 + \lambda_{\max}(S_2)\|e_2\|^2 \tag{48}$$

$$\beta\|e\|^2 \le V_i \le \alpha^2\|e\|^2 \tag{49}$$

where α and β are two positives constants.

Then, Eq. (47) can be written as:

$$\dot{V}_i \leq -(1-\varepsilon)\lambda_1 V_i - \varepsilon\lambda_1 V_i + \lambda_2^{'}\psi\|e\|\forall 0 < \varepsilon < 1 \tag{50}$$

with $\lambda_2^{'} = \alpha\lambda_2$. On the other hand, based on Eq. (49) we can write:

$$-\varepsilon\lambda_1 V_i \leq -\varepsilon\lambda_1\beta\|e\|^2 \tag{51}$$

By substituting (51) into (50), the result will be:

$$\dot{V}_i \leq -(1-\varepsilon)\lambda_1 V_i - \varepsilon\lambda_1^{'}\|e\|^2 + \lambda_2^{'}\psi\|e\| \tag{52}$$

with $\lambda_1^{'} = \beta\lambda_1$.

Then, Eq. (51) can be written as:

$$\dot{V}_i \leq -(1-\varepsilon)\lambda_1 V_i + \|e\|(-\varepsilon\lambda_1^{'}\|e\| + \lambda_2^{'}\psi) \tag{53}$$

$$\dot{V}_i \leq -(1-\varepsilon)\lambda_1 V_i \quad \forall \mathrm{e} \geq \frac{\lambda_2^{\prime}\psi}{\varepsilon\lambda_1^{\prime}} \tag{54}$$

If the machine parameters are known, in this case $\lambda_2 = 0$.So, one just chooses θ_1 and θ_2 such that $\lambda_1 > 0$. If the machine parameters are variable, in this case $\lambda_2 \neq 0$. For $\dot{V}$ defined negative, it is required that the error estimate $\|e\|$ is always greater than $\frac{\lambda_2^{\prime}\psi}{\varepsilon\lambda_1^{\prime}}$. The stability of the error estimation which convergence is arbitrarily set by θ_1 and θ_2 provided that it is always greater than $\frac{\lambda_2\psi}{\varepsilon\lambda_1^{\prime}}$

3.3 Simulations Results and Discussions

The developed observation techniques were applied and tested on a wind energy conversion system using an induction machine. The characteristics of the associated turbine and machine are respectively given in Tables 1 and 2, see Annex.

The suggested presentation of the interconnected observer allows the development of the first observer gain by the fact that state variables do not exist in the computation differential equation loop. This approach guarantees a fast time calculation. As for the adaptive observer, it needs an important calculation time due to the adaptation law. Practically, for that case, an interconnected observer is more efficient than an adaptive observer.

The wind profile used in the simulation is presented as follows Fig. 3.

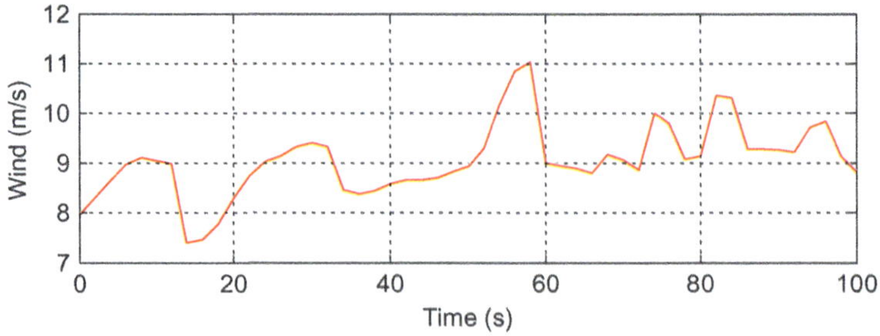

Fig. 3 Wind speed profile

3.3.1 Simulation Results Without Parametric Variation

Figure 4a, b, c and d show the evolution of the mechanical speed and stator flux associated to the WECS with adaptive observer without considering any parametric variation.

Figure 5a, b, c and d show the evolution of mechanical speed and stator flux associated to the WECS with interconnected observer without considering any parametric variation.

Figures 3 and 4 have shown that the estimated and measured mechanical variables of each observer have the same shape with similar amplitude. The estimated magnitudes correctly follow their actual states. In an ideal case, one finds that both methods are relatively accurate.

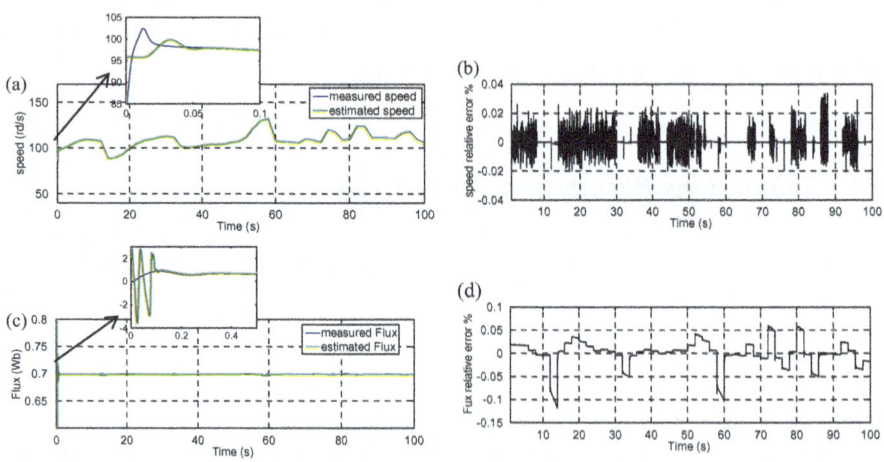

Fig. 4 Adaptive observer simulation results in nominal case: **a**: Evolution of the real and estimated speed, **b**: Evolution of the speed error, **c**: Evolution of the real and estimated flux, **d**: Evolution of the flux error

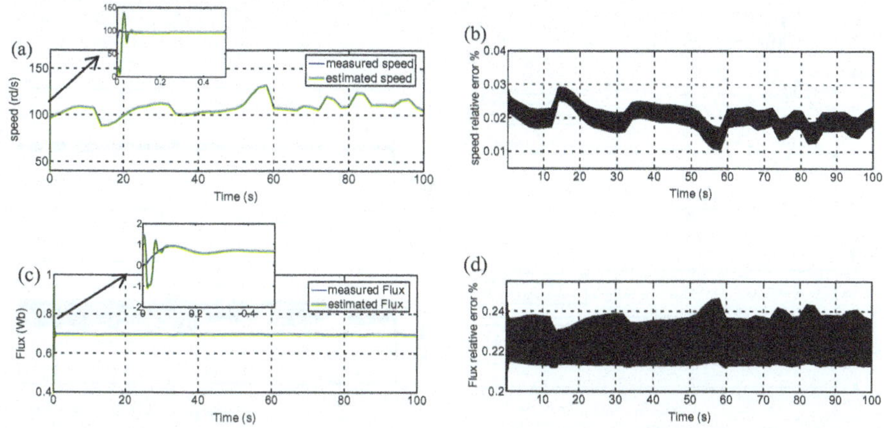

Fig. 5 Interconnected observer simulation results in nominal case: **a**: Evolution of the real and estimated speed, **b**: Evolution of the speed error, **c**: Evolution of the real and estimated flux, **d**: Evolution of the flux error

3.3.2 Simulation Results with Parametric Variation—20% on L_r, L_s and L_m

Considering the inductance variation at around −20%, Fig. 6a, b, c and d show the evolution of the mechanical speed and stator flux associated with the WECS with the adaptive observer.

Figure 7a, b, c and d show the evolution of the mechanical speed and stator flux associated with the WECS with interconnected observers considering the same variations of the IM inductances.

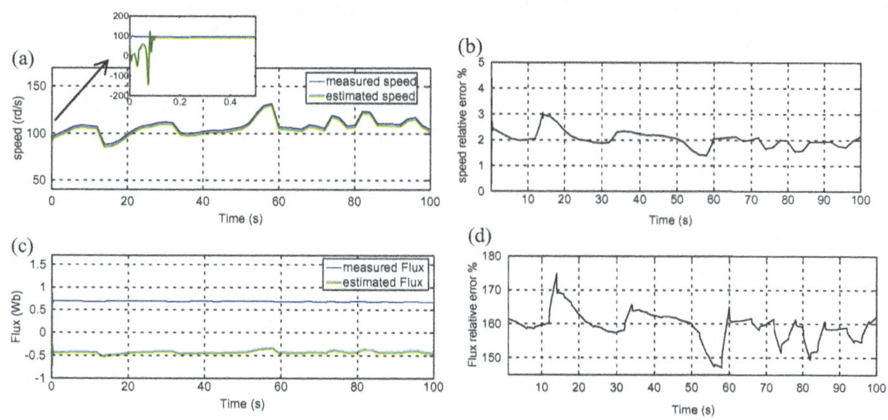

Fig. 6 Adaptive observer simulation results with parametric variation −20% on L_r , L_s and L_m **a**: Evolution of the real and estimated speed, **b**: Evolution of the speed error, **c**: Evolution of the real and estimated flux, **d**: Evolution of the flux error

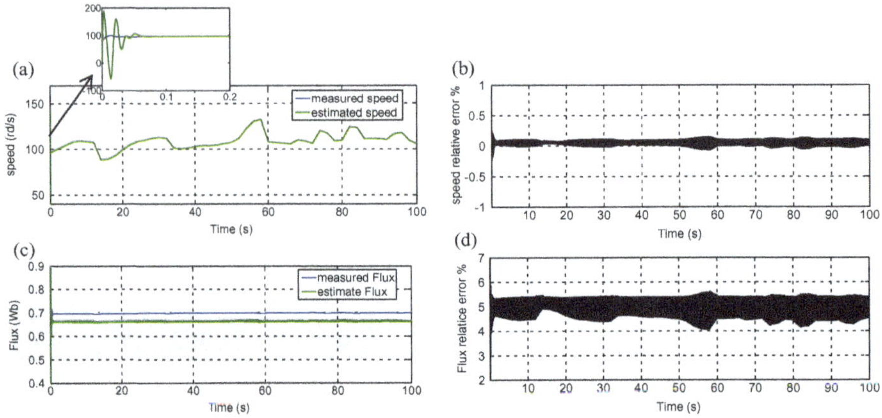

Fig. 7 Interconnected observer simulation results with parametric variation -20% on L_r , L_s and L_m **a**: Evolution of the real and estimated speed, **b**: Evolution of the speed error, **c**: Evolution of the real and estimated flux, **d**: Evolution of the flux error

Following a variation of −20% on inductive parameters, Figs. 5b and 6b show that the interconnected observer presents slight robustness for speed estimation compared to the adaptive observer. While Figs. 5c and 6c prove that interconnected observer is more robust than the adaptive observer. For an eventual control without a mechanical sensor, the estimation of speed and flux are crucial. The adaptive observer presents a major drawback due to the large estimation error of flux under an inductive parametric variation. This implies that the interconnected observer is relatively more robust than the other.

3.3.3 Simulation Results with Parametric Variation + 50% on R_s and R_r

Considering the resistances variation sat around 50%, Fig. 8a, b, c and d show the evolution of the mechanical speed and stator flux associated with the WECS with an adaptive observer Fig. 9a, b, c and d show the evolution of the mechanical speed and stator flux associated with the WECS with an interconnected observer with the same resistances variations.

For a variation of + 50% on the stator and rotor resistances, Figs. 8 and 9 show that both observers present the same degree of robustness.

3.3.4 Wind Torque Simulation Results Using Interconnected Observer

Figure 10 indicate that the interconnected observer estimated wind torque correctly follows its actual state. An advantage of the interconnected observer compared to the adaptive one is that the former allows a direct estimation of the vector state as

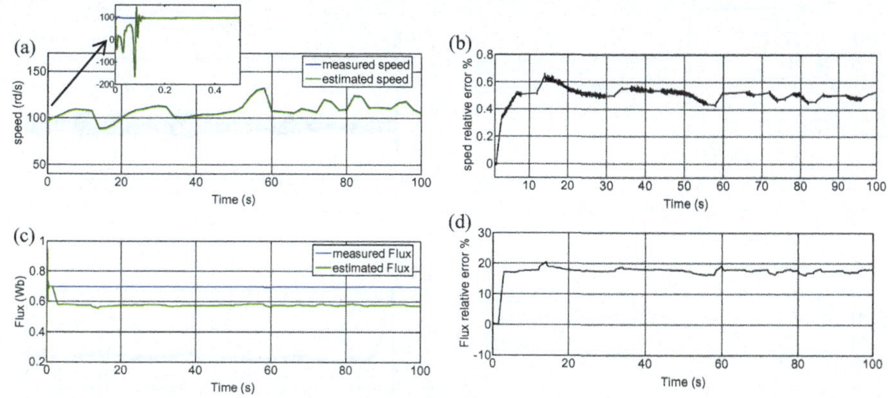

Fig. 8 Adaptive observer simulation results with parametric variation + 50% on R_s and R_r **a**: Evolution of the real and estimated speed, **b**: Evolution of the speed error, **c**: Evolution of the real and estimated flux, **d**: Evolution of the flux error

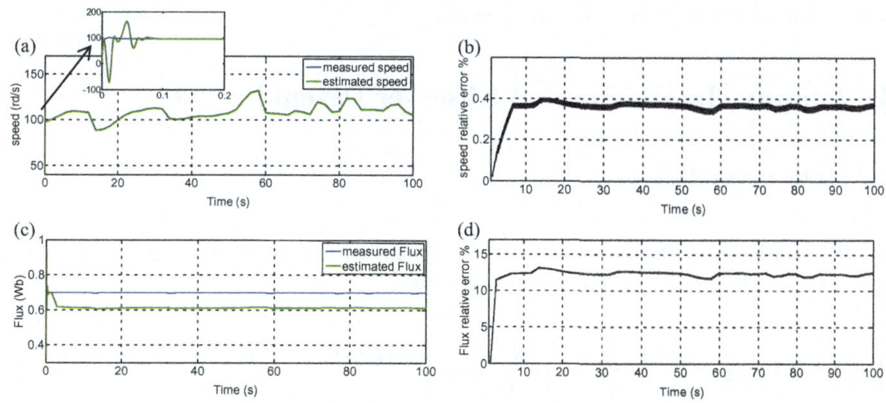

Fig. 9 Interconnected observer simulation results with parametric variation + 50% on R_s and R_r **a**: Evolution of the real and estimated speed, **b**: Evolution of the speed error, **c**: Evolution of the real and estimated flux, **d**: Evolution of the flux error

well as the mechanical magnitudes. The estimated mechanical magnitudes lead to an estimate of the wind torque. While, for the adaptive observer, the mechanical speed is not a state variable in the electric state model. It is considered a parameter. The wind torque cannot be estimated except by an extended Luenberger observer with a stationary assumption of this torque.

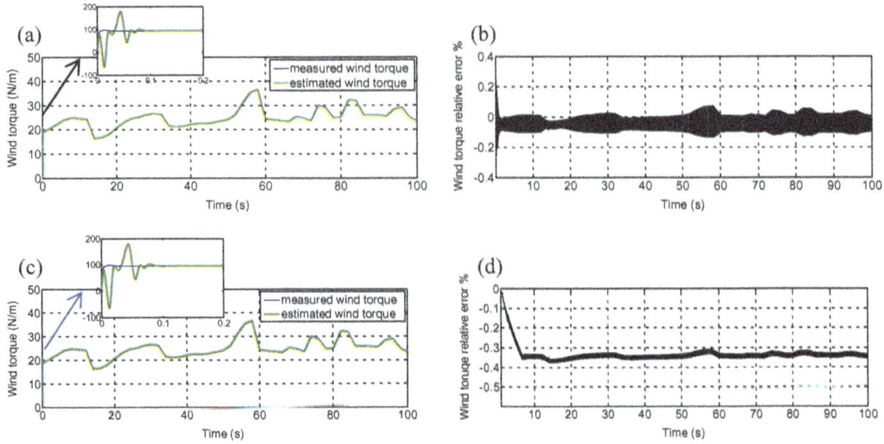

Fig. 10 Interconnected observer simulation results: **a**:Evolution of the real and estimated wind torque with -20% on L_r , L_s and L_m, **b**: Evolution of the wind torque error with -20% on L_r , L_s and L_m, **c**: Evolution of the real and estimated wind torque with $+\ 50\%$ on R_s and R_r, **d**: Evolution of the wind torque error with $+\ 50\%$ on R_s and R_r

4 Adaptive Interconnected Observer Design for IM

4.1 Observer Design

For the synthesis of the observer, we use the following interconnected form which is an extended representation of the model (4):

$$\begin{cases} \frac{d}{dt}\begin{pmatrix} i_{sd} \\ R_s \end{pmatrix} = \begin{pmatrix} -a & -i_{sd}/\sigma L_s \\ 0 & 0 \end{pmatrix}\begin{pmatrix} i_{sd} \\ R_s \end{pmatrix} + \begin{pmatrix} p\Omega\gamma_2\phi_{rq} + \frac{\gamma_2}{\tau_r}\phi_{rd} + \omega_s i_{sq} + m_1 u_{sd} \\ 0 \end{pmatrix} \\ i_{sd} = \begin{pmatrix} 1 & 0 \end{pmatrix}\begin{pmatrix} i_{sd} \\ R_s \end{pmatrix} \end{cases} \tag{55}$$

$$\begin{cases} \frac{d}{dt}\begin{pmatrix} i_{sq} \\ \phi_{rd} \\ \phi_{rq} \end{pmatrix} = \begin{pmatrix} -a & -\gamma_2 p\Omega & \frac{\gamma_2}{\tau_r} \\ 0 & \frac{-1}{\tau_r} & -p\Omega \\ 0 & p\Omega & \frac{-1}{\tau_r} \end{pmatrix}\begin{pmatrix} i_{sq} \\ \phi_{rd} \\ \phi_{rq} \end{pmatrix} + \begin{pmatrix} -\frac{R_s}{\sigma L_s} i_{sq} - \omega_s i_{sd} + m_1 u_{sq} \\ \frac{L_m}{\tau_r} i_{sd} + \omega_s \phi_{rq} \\ \frac{L_m}{\tau_r} i_{sq} - \omega_s \phi_{rd} \end{pmatrix} \\ i_{sq} = \begin{pmatrix} 1 & 0 & 0 \end{pmatrix}\begin{pmatrix} i_{sq} \\ \phi_{rd} \\ \phi_{rq} \end{pmatrix} \end{cases} \tag{56}$$

where $a = \frac{\frac{L_m^2}{L_r^2} R_r}{\sigma L_s}$.

Remark 4 The interconnected representation extends the actual system and it is clear from Eq. (55) that the parameter now acts as a state variable and so the possibility to estimate it through the software sensor. In addition, the variable is considered as a parameter for the two subsystems, its computation was detailed in the stability analysis of the global observer.

The system of Eqs. (55) and (56) can be presented as:

$$\begin{cases} \dot{x}_1 = A_1(y)x_1 + g_1(u, x_2, \Omega) \\ y_1 = C_1 x_1 \end{cases} \tag{57}$$

$$\begin{cases} \dot{x}_2 = A_2(\Omega)x_2 + g_2(u, x_1, x_2) \\ y_2 = C_2 x_2 \end{cases} \tag{58}$$

with:

- For the first subsystem:

$x_1 = [i_{sd} R_s]^T$ is the state vector, $A_1(y) = \begin{pmatrix} -a & -i_{sd}/\sigma L_s \\ 0 & 0 \end{pmatrix}$ is the state matrix, $C_1 = \begin{pmatrix} 1 & 0 \end{pmatrix}$ is the output vector and $g_1(u, x_2, x_1) = \begin{pmatrix} p\Omega\gamma_2\phi_{rq} + \frac{\gamma_2}{\tau_r}\phi_{rd} + \omega_s i_{sq} + m_1 u_{sd} \\ 0 \end{pmatrix}$.

- For the second subsystem:

$x_2 = [i_{sq}\phi_{rd}\phi_{rq}]^T$ is the state vector, $A_2(\Omega) = \begin{pmatrix} -a & -\gamma_2 p\Omega & \frac{\gamma_2}{\tau_r} \\ 0 & \frac{-1}{\tau_r} & -p\Omega \\ 0 & p\Omega & \frac{-1}{\tau_r} \end{pmatrix}$ is the state matrix,

$C_2 = \begin{pmatrix} 1 & 0 & 0 \end{pmatrix}$ is the output vector and $g_2(u, x_1, x_2) = \begin{pmatrix} -\frac{R_s}{\sigma L_s} i_{sq} - \omega_s i_{sd} + m_1 u_{sq} \\ \frac{L_m}{\tau_r} i_{sd} + \omega_s \phi_{rq} \\ \frac{L_m}{\tau_r} i_{sq} - \omega_s \phi_{rd} \end{pmatrix}$.

The two subsystems (57) and (58) are written as affine states, to establish their associated observer.

Assumption 5 State variables of the first subsystem are considered as known inputs for the second subsystem and vice versa.

Assumption 6 It is clear that:

(a) $g_1(u, x_2, \Omega)$ is globally Lipschitz with respect to x_2 uniform to the couple (u, x_1).

(b) $g_2(u, x_1, x_2)$ is globally Lipschitz with respect to x_1 uniform to the couple (u, x_2).

Given these Assumptions, a high gain observer for the system of Eqs. (57) and (58) given respectively by system (59) and (60):

$$\begin{cases} \dot{\hat{x}}_1 = A_1(y)\hat{x}_1 + g_1\left(u, \hat{x}_2, \hat{\Omega}\right) + M_1(\theta_1)(y_1 - \hat{y}_1) \\ \hat{y}_1 = C_1\hat{x}_1 \end{cases} \tag{59}$$

$$\begin{cases} \dot{\hat{x}}_2 = A_2\left(\hat{\Omega}\right)\hat{x}_2 + g_2\left(u, \hat{x}_1, \hat{x}_2\right) + M_2(\hat{\Omega}, \theta_2)(y_2 - \hat{y}_2) \\ \hat{y}_2 = C_2\hat{x}_2 \end{cases} \tag{60}$$

with.

- For the first observer:

$M_1(\theta_1, y) = S_1^{-1}C_1^T$ is the observer gain where S_1 is a defined positive symmetric matrix solution of the differential equation defined by[10]:

$$\dot{S}_1(\theta_1, y) = -\theta_1 S_1(\theta_1, y) - A_1^T S_1(\theta_1, y) - S_1(\theta_1, y)A_1 + C_1^T C_1 \tag{61}$$

- For the second observer:

$M_2(\hat{\Omega}, \theta_2) = S_2^{-1}C_2^T$ is the gain of the observer where S_2 is a defined positive symmetric matrix solution of the differential equation defined by [10]:

$$\dot{S}_2(\hat{\Omega}, \theta_2) = -\theta_2 S_2(\hat{\Omega}, \theta_2) - A_2^T S_2(\hat{\Omega}, \theta_2) - S_2(\hat{\Omega}, \theta_2)A_2 + C_2^T C_2 \tag{62}$$

where θ_1 and θ_2 are positive constants.

The figure below presents the block diagram of the adaptive interconnected observer. The adaptation law is determined from the stability analysis of the high gains interconnected observer Fig. 11.

4.2 *Stability Analysis*

To prove the stability of the hybrid observer, consider the estimation error defined by:

$$e_1 = x_1 - \hat{x}_1;\ e_2 = x_2 - \hat{x}_2 \tag{63}$$

The estimation error dynamics e_1 is defined by:

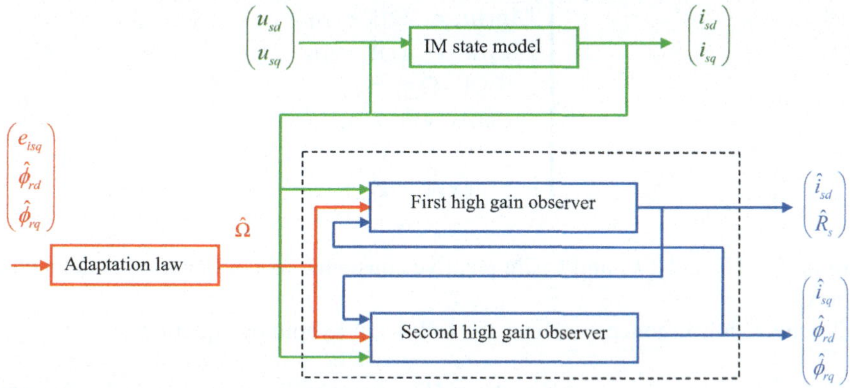

Fig. 11 Block diagram of the adaptive interconnected observer

$$\dot{e}_1 = [A_1(y) - S_1^{-1}C_1^T C_1]e_1 + g_1(u, x_2, \Omega) - g_1\left(u, \hat{x}_2, \hat{\Omega}\right) \tag{64}$$

The dynamics of the estimation error e_2 is defined by:

$$\dot{e}_2 = [A_2\left(\hat{\Omega}\right) - S_2^{-1}C_2^T C_2]e_2 + g_2(u, x_2, \Omega) - g_1\left(u, \hat{x}_1, \hat{\Omega}\right) + \left[A_2(\Omega) - A_2\left(\hat{\Omega}\right)\right]x_2 \tag{65}$$

Now consider the Eqs. (64) and (65) with uncertainties on the parameters:

$$\dot{e}_1 = [A_1(y) - S_1^{-1}C_1^T C_1]e_1 + \Delta g_1 + \delta g_1 + \delta A_1(y)e_1 \tag{66}$$

$$\dot{e}_2 = [A_2\left(\hat{\Omega}\right) - S_2^{-1}C_2^T C_2]e_2 + \Delta g_2 + \Delta A_2 x_2 + \delta g_2 + \delta A_2 x_2 \tag{67}$$

where $\Delta g_1 = g_1(u, x_2, \Omega) - g_1\left(u, \hat{x}_2, \hat{\Omega}\right)$,$\Delta g_2 = g_2(u, x_1, x_2) - g_2\left(u, \hat{x}_1, \hat{x}_2\right)$,$\Delta A_2 = A_2(\Omega) - A_2(\hat{\Omega})\delta g_1(u, x_2, \Omega)$, and $\delta g_2(u, x_2, \Omega)$,$\delta A_1(x_1)$ and $\delta A_2(\Omega)$ are respectively the uncertain terms of $g_1(u, x_2, \Omega)$,$g_2(u, x_1, x_2)$,$A_1(y)$ and $A_2(\Omega)$.

The machine parameters are known with some precision and they are bounded. Taking the initial conditions of the IM drive and the observer in the physical operation domain D_2 and by the fact that the uncertainties are bounded, the following inequalities hold:

$$\begin{cases} \|\delta g_1(u, x_2, \Omega)\| < \rho_1 \\ \|\delta g_2(u, x_2, x_1)\| < \rho_2 \\ \|\delta A_2(\Omega)\| < \rho_3 \\ \|\delta A_1(y)\| < \rho_4 \\ \|x_1\| < k_4 \\ \|x_2\| < k_5 \end{cases} \tag{68}$$

with $\rho_i, i \in \{1, .., 4\}$, k_4 and k_5 are positive constants, its computation is detailed in Annex.

To check convergence, consider the following Lyapunov equation:

$$V = V_1 + V_2 + \lambda(\Delta\Omega)^2 \tag{69}$$

where: $V_1 = e_1^T S_1 e_1$, $V_2 = e_2^T S_2 e_2$, $\Delta\Omega = \Omega - \hat{\Omega}$ and λ is a positive constant.

Calculating the derivative of V along trajectories of e_1 and e_2 we obtain:

$$\begin{aligned} \dot{V} &= e_1^T[(A_1(y) - S_1^{-1}C_1^T C_1)^T S_1 + S_1(A_1(y) - S_1^{-1}C_1^T C_1) + \dot{S}_1]e_1 \\ &+2e_1^T S_1[\Delta g_1 + \delta g_1] + 2e_1^T S_1 \delta A_1 e_1 \\ &+e_2^T\left[\left(A_2(\hat{\Omega}) - S_2^{-1}C_2^T C_2\right)^T S_2 + S_2\left(A_2(\hat{\Omega}) - S_2^{-1}C_2^T C_2\right) + \dot{S}_2\right]e_2 \\ &+2e_2^T S_2[\Delta A_2 + \delta A_2]x_2 + 2e_2^T S_2[\Delta g_2 + \delta g_2] + 2\lambda\Delta\Omega\frac{d}{dt}\hat{\Omega} \end{aligned} \tag{70}$$

The function $\dot{V}$ is decomposed in two functions σ_1 and σ_2 with σ_1 defined by:

$$\begin{aligned} \sigma_1 &= e_1^T[(A_1(y) - S_1^{-1}C_1^T C_1)^T S_1 + S_1(A_1(y) - S_1^{-1}C_1^T C_1) + \dot{S}_1]e_1 \\ &+e_2^T\left[\left(A_2(\hat{\Omega}) - S_2^{-1}C_2^T C_2\right)^T S_2 + S_2\left(A_2(\hat{\Omega}) - S_2^{-1}C_2^T C_2\right) + \dot{S}_2\right]e_2 \\ &+2e_1^T S_1[\Delta g_1 + \delta g_1] + 2e_2^T S_2[\Delta g_2 + \delta g_2] + 2e_1^T S_1 \delta A_1 e_1 + 2e_2^T S_2 \delta A_2(\Omega) x_2 \end{aligned} \tag{71}$$

and σ_2 is defined by:

$$\sigma_2 = 2e_2^T S_2 \Delta A_2 x_2 + 2\lambda\Delta\Omega\frac{d}{dt}\hat{\Omega} \tag{72}$$

The observer stability is proven if the derivative of the Lyapunov equation is negative. If we consider $\sigma_1 < 0$ and $\sigma_2 = 0$, the stability condition of the observer is proved.

Introducing standards for the function σ_1, we can write:

$$\begin{aligned}\sigma_1 &\leq -(\theta_1 - 2\rho_4)e_1^T S_1 e_1 - \theta_2 e_2^T S_2 e_2 \\ &+2\|e_1\|\|S_1\|\|\Delta g_1 + \delta g_1\| + 2\|e_2\|\|S_2\|\|\Delta g_2 + \delta g_2\| \\ &+2\|e_2\|\|S_2\|\|\delta A_2(\Omega)\|\|x_2\|\end{aligned} \tag{73}$$

It is well known that: $\forall a, b \in \mathbb{R} a + b \leq a + b$, so:

$$\begin{aligned}\sigma_1 &\leq -(\theta_1 - 2\rho_4)e_1^T S_1 e_1 - \theta_2 e_2^T S_2 e_2 \\ &+2\|e_1\|\|S_1\|\|\Delta g_1\| + 2\|e_1\|\|S_1\|\|\delta g_1\| \\ &+2\|e_2\|\|S_2\|\|\Delta g_2\| + 2\|e_2\|\|S_2\|\|\delta g_2\| \\ &+2\|e_2\|\|S_2\|\|\delta A_2(\Omega)\|\|x_2\|\end{aligned} \tag{74}$$

Knowing that g_1,g_2 and A_1 are globally Lipschitz, one can write:

$$\begin{cases}\|\Delta g_1\| \leq k_1\|e_2\| \\ \|\Delta g_2\| \leq k_2\|e_1\|\end{cases} \tag{75}$$

where k_1 and k_2 are positive constants, its computation is detailed in Annex.

Assume that the known inputs of each observer are regularly persistent. According to the Lemma 1, there exist real numbers $\lambda_{min}(S_1) > 0$,$\lambda_{\max}(S_1) > 0$,$\lambda_{min}(S_2) > 0$,$\lambda_{\max}(S_2) > 0$,$\upsilon_1 > 0$ and $\upsilon_2 > 0$ such that:

$$\begin{cases}\|S_1\| \leq \upsilon_1 \\ \|S_2\| \leq \upsilon_2 \\ \lambda_{min}(S_1)\|e_1\|^2 \leq V_1 \leq \lambda_{max}(S_1)\|e_1\|^2 \\ \lambda_{min}(S_2)\|e_2\|^2 \leq V_2 \leq \lambda_{max}(S_2)\|e_2\|^2\end{cases} \tag{76}$$

The condition on the function σ_1 becomes:

$$\sigma_1 \leq -(\theta_1 - 2\rho_4)V_1 - \theta_2 V_2 + \delta\sqrt{V_1}\sqrt{V_2} + \mu_1\|e_1\| + \mu_2\|e_2\| \tag{77}$$

where $\delta = \frac{2k_1\upsilon_1 + 2k_2\upsilon_2}{\sqrt{\lambda_{min}(S_1)}\sqrt{\lambda_{min}(S_2)}}$, $\mu_1 = 2\upsilon_1\rho_1$ and $\mu_2 = 2\upsilon_2(\rho_2 + \rho_3 k_5)$.

On the other hand, taking into account Eq. (46), the condition on the function σ_1 defined on (77) becomes:

$$\sigma_1 \leq -\left(\theta_1 - 2\rho_4 - \frac{\delta}{2}\right)V_1 - \left(\theta_2 - \frac{\delta}{2}\right)V_2 + \mu_1\|e_1\| + \mu_2\|e_2\| \tag{78}$$

$$\sigma_1 \leq -\left(\theta_1 - 2\rho_4 - \frac{\delta}{2}\right)V_1 - \left(\theta_2 - \frac{\delta}{2}\right)V_2 + \frac{\mu_1}{\sqrt{\lambda_{min}(S_1)}}\sqrt{V_1} + \frac{\mu_2}{\sqrt{\lambda_{min}(S_2)}}\sqrt{V_2} \tag{79}$$

We suppose that $\lambda_1 = \min(\theta_1 - 2\rho_4 - \frac{\delta}{2}, \theta_2 - \frac{\delta}{2})$ and $\lambda_2 = \max(\frac{\mu_1}{\sqrt{\lambda_{min}(S_1)}}, \frac{\mu_2}{\sqrt{\lambda_{min}(S_2)}})$.

Then,

$$\sigma_1 \leq -\lambda_1(V_1 + V_2) + \lambda_2(\sqrt{V_1} + \sqrt{V_2}) \tag{80}$$

That, if we consider the function $H = V_1 + V_2$, we can write:

$$\sigma_1 \leq -\lambda_1 H + \lambda_2\sqrt{H} \tag{81}$$

Taking into account (76) we can write:

$$\lambda_{\min}(S_1)\|e_1\|^2 + \lambda_{\min}(S_2)\|e_2\|^2 \leq H \leq \lambda_{\max}(S_1)\|e_1\|^2 + \lambda_{\max}(S_2)\|e_2\|^2 \tag{82}$$

$$\beta\|e\|^2 \leq H \leq \alpha^2\|e\|^2 \tag{83}$$

where α and β are two positives constants.

Then, Eq. (81) can be written as:

$$\sigma_1 \leq -(1-\varepsilon)\lambda_1 H - \varepsilon\lambda_1 H_i + \lambda_2^{'}\|e\|\forall 0 < \varepsilon < 1 \tag{84}$$

with $\lambda_2^{'} = \alpha\lambda_2$. On the other hand, based on Eq. (49), we can write:

$$-\varepsilon\lambda_1 H \leq -\varepsilon\lambda_1\beta\|e\|^2 \tag{85}$$

By substituting (85) into (84), the result will be:

$$\sigma_1 \leq -(1-\varepsilon)\lambda_1 H - \varepsilon\lambda_1^{'}\|e\|^2 + \lambda_2^{'}\|e\| \tag{86}$$

with $\lambda_1^{'} = \beta\lambda_1$.

Then, Eq. (56) can be written as:

$$\sigma_1 \leq -(1-\varepsilon)\lambda_1 H + \|e\|(-\varepsilon\lambda_1^{'}\|e\| + \lambda_2^{'}) \tag{87}$$

$$\sigma_1 \leq -(1-\varepsilon)\lambda_1 H \quad \forall \mathrm{e} \geq \frac{\lambda_2^{'}}{\varepsilon\lambda_1^{'}} \tag{88}$$

If the machine parameters are known, in this case $\lambda_2 = 0$. To check the first condition, we just choose θ_1 and θ_2 such that $\lambda_1 > 0$. If the machine parameters are variables, in this case $\lambda_2 \neq 0$. For σ_1 defined negative, it is required that the error estimate $\|e\|$ is always greater than $\frac{\lambda_2^{'}}{\varepsilon\lambda_1^{'}}$ The stability of the error estimation which convergence is arbitrarily set by θ_1 and.

Considering the validation of the first condition, to ensure the observer stability, we have to admit that σ_2 is equal to zero , then we obtain:

$$\frac{d\hat{\Omega}}{dt} = -\frac{p}{\lambda} e_2^T S_2 A_0 x_2 \tag{90}$$

with: $A_0 = \begin{pmatrix} 0 & -\gamma_2 & 0 \\ 0 & 0 & -1 \\ 0 & 1 & 0 \end{pmatrix}$

The adaptation law$\hat{\Omega}$ is defined by:

$$\frac{d\hat{\Omega}}{dt} = -\frac{p}{\lambda}[e_{isq}(a\phi_{rd} - S_{12}\phi_{rq}) + e_{\phi rd}(b\phi_{rd} - S_{22}\phi_{rq}) + e_{\phi rq}(c\phi_{rd} - S_{23}\phi_{rq})] \tag{91}$$

where $a = -\gamma_2 S_{11} + S_{13}$, $b = -\gamma_2 S_{12} + S_{23}$ and $c = -\gamma_2 S_{13} + S_{33}$ with S_{11}, S_{12}, S_{13}, S_{22}, S_{23} and S_{33} are the terms of the matrix S_2 such that $S_2 = \begin{pmatrix} S_{11} & S_{12} & S_{13} \\ S_{12} & S_{22} & S_{23} \\ S_{13} & S_{23} & S_{33} \end{pmatrix}$.

Knowing that $e_{\phi rd} = \phi_{rd} - \hat{\phi}_{rd}$ Eq. (40) becomes:

$$\frac{d\hat{\Omega}}{dt} = f_1(x, \hat{x}) + f_2(\phi, \hat{\phi}) \tag{92}$$

where

$$f_1(x, \hat{x}) = -\frac{p}{\lambda} e_{isq}\left(a\hat{\phi}_{rd} - S_{12}\hat{\phi}_{rq}\right)$$

and

$$f_2(\phi, \hat{\phi}) = -\frac{p}{\lambda}[e_{isq}(ae_{\phi rd} - S_{12}e_{\phi rq}) + e_{\phi rd}(b\phi_{rd} - S_{22}\phi_{rq}) + e_{\phi rq}(c\phi_{rd} - S_{23}\phi_{rq})]$$

4.3 Wind Torque Observer

The wind torque T_w is estimated by an extended Luenberger observer. This deterministic approach is used to estimate variables or parameters of the machine as a separate part of its state variables. The estimate can then be extended to the wind torque or some electrical parameters assuming a very low variation of these quantities. It is assumed that:

$$\frac{d}{dt} T_w = 0 \tag{93}$$

One constructs the increased estimated mechanical magnitude state model defined by:

$$\begin{cases} \frac{d}{dt}X_e = A_e X_e + B_e \hat{T}_{em} \\ \hat{\Omega} = C_e X_e \end{cases} \tag{94}$$

with $X_e = \begin{bmatrix} \Omega \\ T_w \end{bmatrix}$, $A_e = \begin{bmatrix} -\frac{f_v}{J} & \frac{1}{J} \\ 0 & 0 \end{bmatrix}$, $B_e = \begin{bmatrix} -\frac{1}{J} \\ 0 \end{bmatrix}$ and $C_e = \begin{bmatrix} 1 \; 0 \end{bmatrix}$.

The induction machine electromagnetic torque is calculated from the estimated electrical magnitude such that:

$$\hat{T}_{em} = p\frac{L_m}{L_r}(\hat{i}_{sq}\hat{\phi}_{rd} - \hat{i}_{sd}\hat{\phi}_{rq}) \tag{95}$$

The equations system (94) is a linear state model to which we can associate the extended Luenberger observer as follows:

$$\frac{d}{dt}\hat{X}_e = A_e\hat{X}_e + B_e\hat{T}_{em} + L(\hat{\Omega} - \Omega_{est}) \tag{96}$$

where $\hat{X}_e = \begin{bmatrix} \Omega_{est} \\ \hat{T}_w \end{bmatrix}$, G is the observer gain defined by $L = \begin{bmatrix} l_1 \\ l_2 \end{bmatrix}$.

Figure 12 illustrate the block diagram of the developed extended Luenberger observer.

The stability of the observer will be ensured by an appropriate choice of the gain L so that $(A_e - LC_e)$ is stable, see Annex Fig. 13.

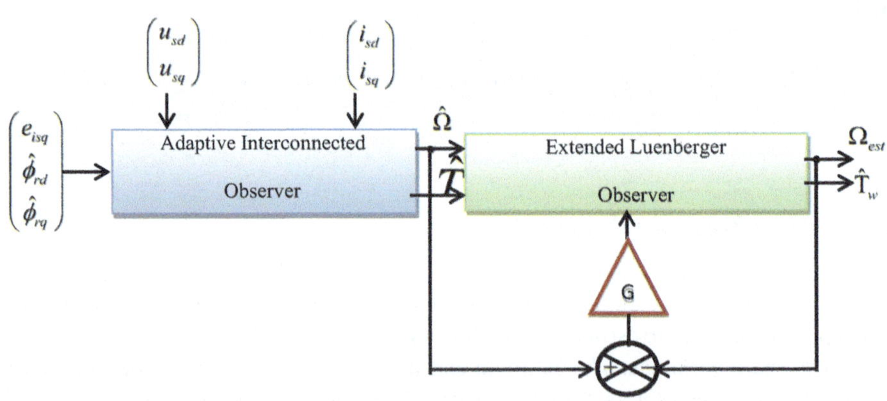

Fig. 12 Block diagram of the extended Luenberger observer

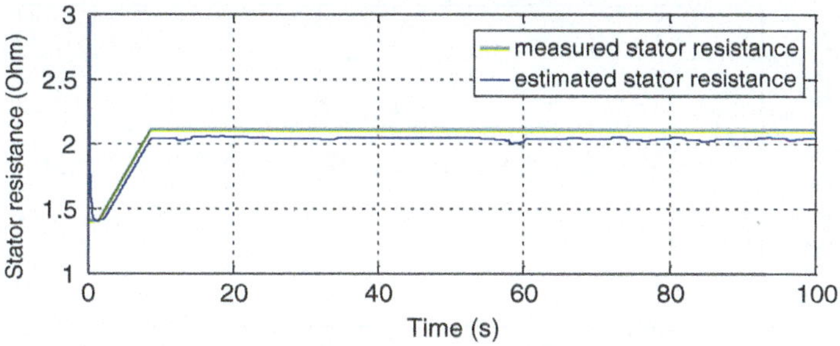

Fig. 13 Evolution of the IM stator resistance and the adaptive interconnected estimated stator resistance with +50% variation R_r on and R_s

Fig. 14 Evolution of the induction machine speed and the adaptive interconnected observer estimated speed with +50% variation on R_r and R_s

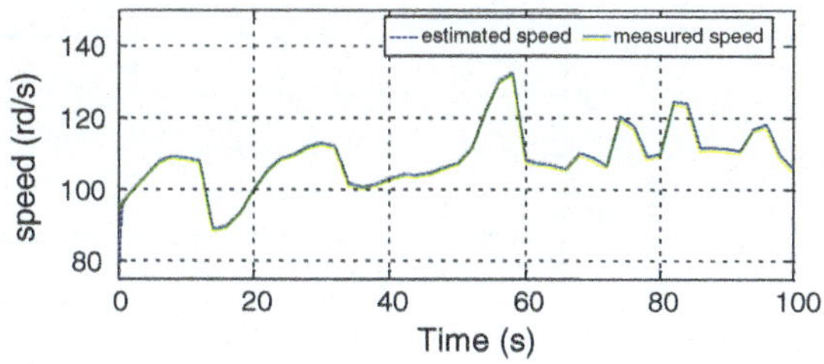

4.4 *Simulation Results*

The characteristics of the associated turbine and machine are respectively given in Tables 1 and 2. In this study, the variation is tested with different parameters: namely the resistances (R_r and R_s) and the inductances (L_r and L_m). The wind profile used in the simulation is presented in Fig. 2.

4.4.1 Simulation Results with +50% Variation on R_r and R_s

Despite the parametric variation +50% on R_r and R_s, the proposed observer converges and gives desirable results. It is clear from Figs. 14, 15, 16 and 17 that the estimated states follow their actual variables with an average of a relative speed error around 0.2% Fig. 18.

4.4.2 Simulation Results with + 50% Variation on R_s and −20% on L_r and L_m

Figure 20 it is clear from Figs 19 and 21 that the estimated and measurable magnitudes have similar shapes and amplitudes. A variation of about -20% on the inductances and +50% on the stator resistance maintains the performance of the observer. Besides,

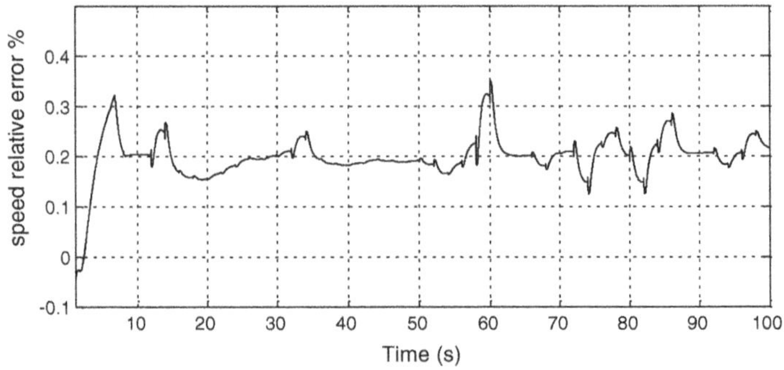

Fig. 15 Evolution of the adaptive interconnected observer speed relative error with +50% variation on R_r and R_s

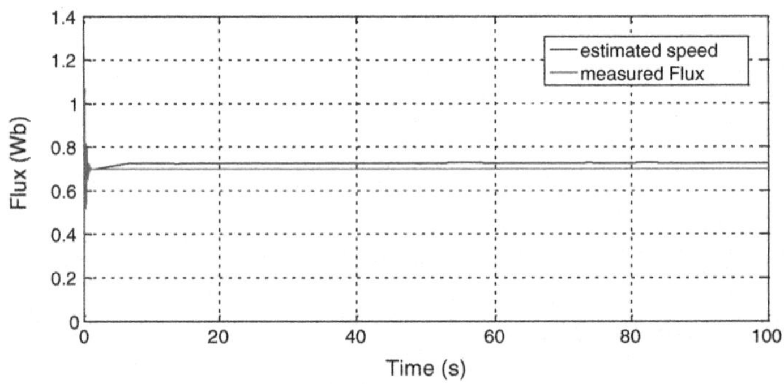

Fig. 16 Evolution of the induction machine Flux and the adaptive interconnected observer estimated Flux with +50% variation on R_r and R_s

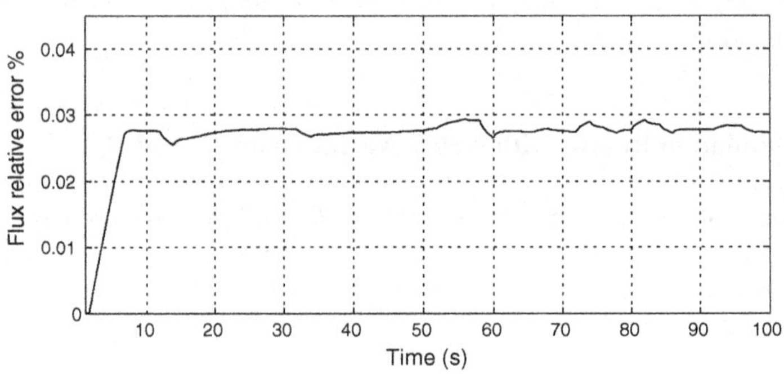

Fig. 17 Evolution of the adaptive interconnected observer Flux relative error with +50% variation on R_r and R_s

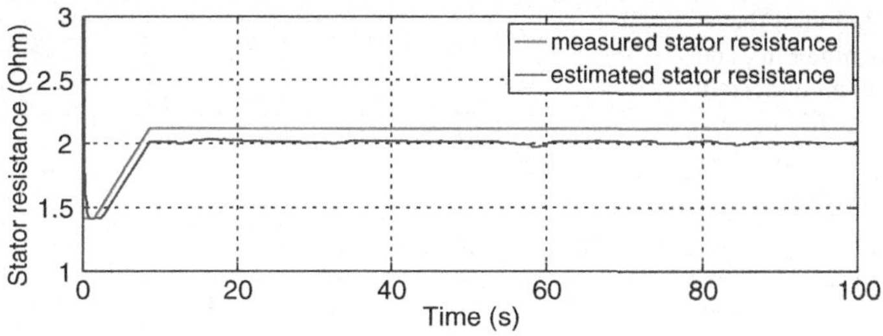

Fig. 18 Evolution of the IM stator resistance and the adaptive interconnected estimated stator resistance with +50% variation on R_s and –20% on L_r and L_m

Fig. 22 show an average flux relative error of around 0.1%. As a consequence, the speed estimation keeps its performance with an average speed relative error of around 0.5%. The advantage of the stator resistance estimation has become obvious. Figure 22 has shown the evolution of the estimated induction machine wind torque compared to the real one. According to Fig. 23, it can be deduced that the extended Luenberger observer converge and gives a desirable result.

Fig. 19 Evolution of the induction machine speed and the adaptive interconnected observer estimated speed with +50% variation on R_s and –20% on L_r and L_m

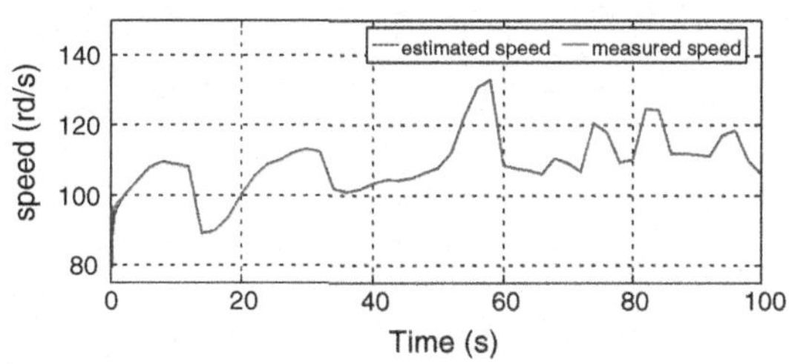

Fig. 20 Evolution of the adaptive interconnected observer speed relative error with +50% variation on R_s and –20% on L_r and L_m

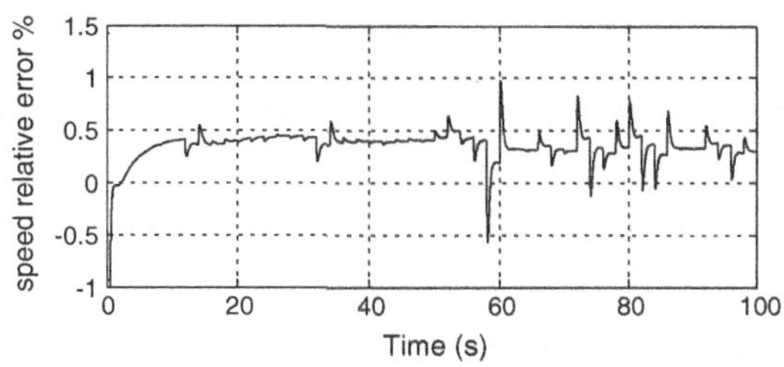

Fig. 21 Evolution of the induction machine Flux and the adaptive interconnected observer estimated Flux with +50% variation on R_s and -20% on L_r and L_m

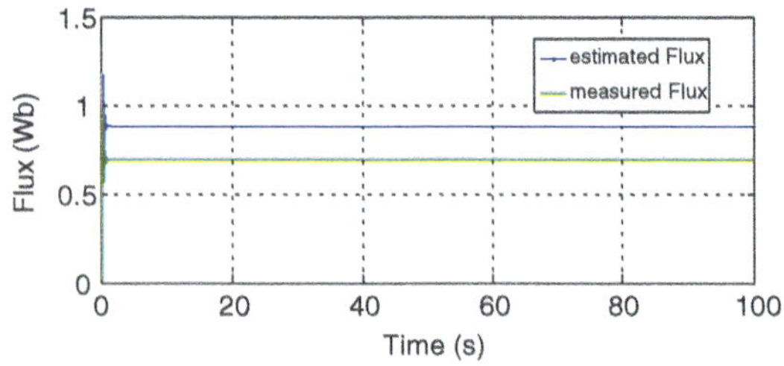

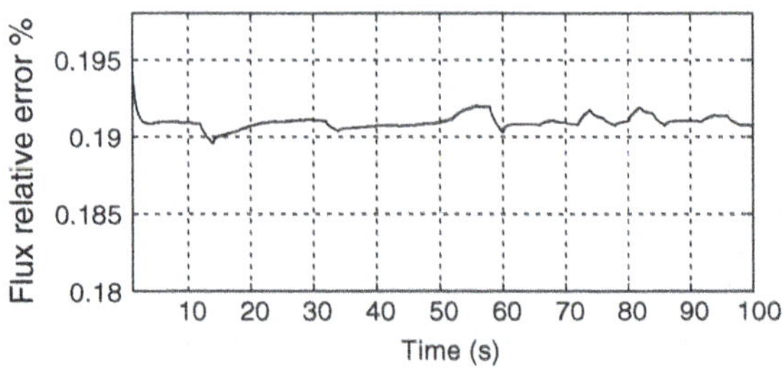

Fig. 22 Evolution of the adaptive interconnected observer Flux relative error with +50% variation on R_s and -20% on L_r and L_m

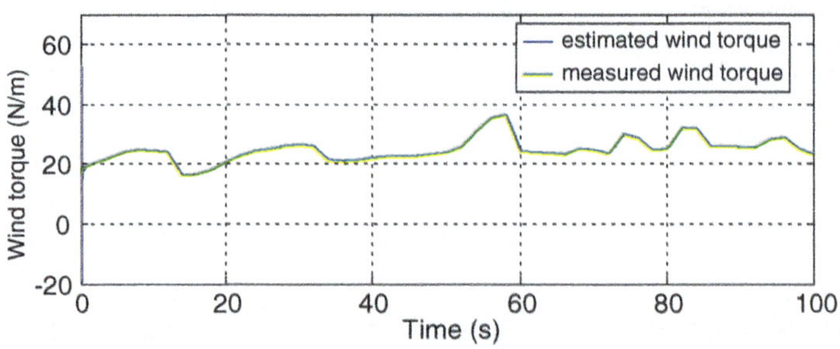

Fig. 23 Evolution of the estimated induction machine wind torque and the actual wind torque

5 Adaptive Interconnected Observer Design for PMSM

5.1 Observer Design

For the synthesis of the observer, we use the following interconnected form which is an extended representation of the model (7):

$$\begin{cases} \frac{d}{dt}\begin{pmatrix} i_d \\ L_s^{-1} \end{pmatrix} = \begin{pmatrix} 0 & -R_s i_d \\ 0 & 0 \end{pmatrix}\begin{pmatrix} i_d \\ L_s^{-1} \end{pmatrix} + \begin{pmatrix} p\Omega i_q \\ 0 \end{pmatrix} + \begin{pmatrix} L_s^{-1} v_d \\ 0 \end{pmatrix} \\ i_d = \begin{pmatrix} 1 & 0 \end{pmatrix}\begin{pmatrix} i_d \\ L_s^{-1} \end{pmatrix} \end{cases} \tag{97}$$

$$\begin{cases} \frac{d}{dt}\begin{pmatrix} i_q \\ R_s \end{pmatrix} = \begin{pmatrix} 0 & -L_s^{-1} i_q \\ 0 & 0 \end{pmatrix}\begin{pmatrix} i_q \\ R_s \end{pmatrix} + \begin{pmatrix} -p\Omega i_d \\ 0 \end{pmatrix} + \begin{pmatrix} -p\phi_f L_s^{-1}\Omega + L_s^{-1} v_q \\ 0 \end{pmatrix} \\ i_q = \begin{pmatrix} 1 & 0 \end{pmatrix}\begin{pmatrix} i_q \\ R_s \end{pmatrix} \end{cases} \tag{98}$$

The variable Ω is considered as a parameter for the two subsystems, Its computation will be detailed in the stability analysis of the observer. The system of Eqs. (97) and (98) can be presented as:

$$\begin{cases} \dot{x}_1 = A_1(x_2, y)x_1 + g_1(\Omega, y) + \Phi_1(x_1, u) \\ y_1 = C_1 x_1 \end{cases} \tag{99}$$

$$\begin{cases} \dot{x}_2 = A_2(x_1, y)x_2 + g_2(\Omega, y) + \Phi_2(x_1, \Omega, u) \\ y_2 = C_2 x_2 \end{cases} \tag{100}$$

with:

For the first subsystem:

$x_1 = [i_d L_s^{-1}]^T$ is the state vector, $A_1(x_2, y) = \begin{pmatrix} 0 & -R_s i_d \\ 0 & 0 \end{pmatrix}$ is the state matrix, $C_1 = \begin{pmatrix} 1 & 0 \end{pmatrix}$ is the output vector, y is the output term, $g_1(\Omega, y) = \begin{pmatrix} p\Omega i_q \\ 0 \end{pmatrix}$ and $\Phi_1(x_1, u) = \begin{pmatrix} L_s^{-1} v_d \\ 0 \end{pmatrix}$.

For the second subsystem:

$x_2 = [i_q R_s]^T$ is the state vector, $A_2(x_1, y) = \begin{pmatrix} 0 & -L_s^{-1} i_q \\ 0 & 0 \end{pmatrix}$ is the state matrix, $C_2 = \begin{pmatrix} 1 & 0 \end{pmatrix}$ is the output vector, y_2 is the output term, $g_2(\Omega, y) = \begin{pmatrix} -p\Omega i_d \\ 0 \end{pmatrix}$ and $\Phi_2(x_1, \Omega, u) = \begin{pmatrix} -p\phi_f L_s^{-1}\Omega + L_s^{-1} v_q \\ 0 \end{pmatrix}$.

The two subsystems (50) and (51) are written as affine state presentation.

Assumption 7: State variables of the first subsystem are considered as known inputs for the second subsystem and vice versa.

Assumption 8: We have as an assumption the following four points:

(a) $A_1(x_2, y)$ is globally Lipschitz with respect to x_2 and uniformly with respect to y.
(b) $A_2(x_1, y)$ is globally Lipschitz with respect to x_1 and uniformly with respect to y.
(c) $\Phi_1(x_1, u)$ is globally Lipschitz with respect to x_1.
(d) $\Phi_2(x_1, \Omega, u)$ is globally Lipschitz with respect to x_1 and uniformly with respect to (Ω, u).

Given these remarks, we associate with the systems (99) and (100) respectively the high gains observers (101) and (102):

$$\begin{cases} \dot{\hat{x}}_1 = A_1(\hat{x}_2, y)\hat{x}_1 + g_1(\hat{\Omega}, y) + \Phi_1(\hat{x}_1, u) + S_1^{-1}(\theta_1, \hat{x}_2)C_1^T(y_1 - \hat{y}_1) \\ \hat{y}_1 = C_1\hat{x}_1 \end{cases} \tag{101}$$

$$\begin{cases} \dot{\hat{x}}_2 = A_2(\hat{x}_1, y)\hat{x}_2 + g_2(\hat{\Omega}, y) + \Phi_2(\hat{x}_1, \hat{\Omega}, u) + S_2^{-1}(\theta_2, \hat{x}_1)C_2^T(y_2 - \hat{y}_2) \\ \hat{y}_2 = C_2\hat{x}_2 \end{cases} \tag{102}$$

with.

- For the first observer:

$\hat{x}_1$ is the estimated vector of x_1,$A_1(\hat{x}_2, y)$ is the estimated matrix of $A_1(x_2, y)$,$g_1(\hat{\Omega}, y)$ is the estimated vector of $g_1(\Omega, y)$, $\Phi_1(\hat{x}_1, u)$ is the estimated vector of $\Phi_1(x_1, u)$ and $S_1^{-1}(\theta_1, \hat{x}_2)C_1^T$ is the observer gain where S_1 is a defined positive symmetric matrix solution of the differential equation defined by [10]:

$$\dot{S}_1(\theta_1, \hat{x}_2) = -\theta_1 S_1(\theta_1, \hat{x}_2) - A_1^T S_1(\theta_1, \hat{x}_2) - S_1(\theta_1, \hat{x}_2)A_1 + C_1^T C_1 \tag{103}$$

with θ_1 is a positive constant.

- For the second observer:

$\hat{x}_2$ is the estimated vector of x_2,$A_2(\hat{x}_1, y)$ is the estimated matrix of $A_2(x_1, y)$,$g_2(\hat{\Omega}, y)$ is the estimated vector of $g_2(\Omega, y)$, $\Phi_2(\hat{x}_1, \hat{\Omega}, u)$ is the estimated vector of $\Phi_2(x_1, \Omega, u)$ and $S_2^{-1}(\theta_2, \hat{x}_1)C_2^T$ is the observer gain where S_2 is a defined positive symmetric matrix solution of the differential equation defined by[10]:

$$\dot{S}_2(\theta_2, \hat{x}_1) = -\theta_2 S_2(\theta_2, \hat{x}_1) - A_2^T S_2(\theta_2, \hat{x}_1) - S_2(\theta_2, \hat{x}_1)A_2 + C_2^T C_2 \tag{104}$$

with θ_2 is a positive constant.

The figure below presents the block diagram of the adaptive interconnected observer. The adaptation law is determined from the stability analysis of the high gains interconnected observer Fig. 24.

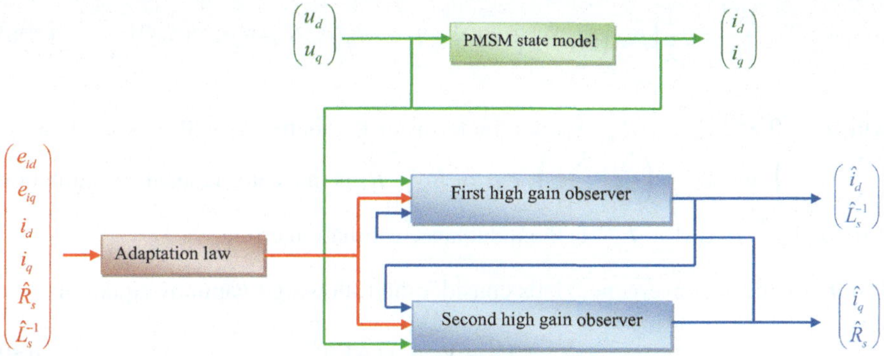

Fig. 24 Block diagram of the adaptive interconnected observer for. PMSM

The terms e_{id} and e_{iq} in Fig. 1 correspond to the current errors with $e_{id} = i_d - \hat{i}_d$ and $e_{iq} = i_q - \hat{i}_q$. $\hat{i}_d$,$\hat{i}_q$, $\hat{R}_s$, $\hat{L}_s^{-1}$ and $\hat{\Omega}$ are respectively the estimated magnitude of i_d,i_q, R_s, L_s^{-1} and Ω.

5.2 *Stability Analysis*

To prove the stability of the hybrid observer, consider the estimation error defined by:

$$e_1 = x_1 - \hat{x}_1; e_2 = x_2 - \hat{x}_2 \tag{105}$$

The dynamics of the estimation error e_1 and e_2 are defined by:

$$\dot{e}_1 = [A_1(\hat{x}_2, y) - S_1^{-1}C_1^T C_1]e_1 + \Delta A_1 x_1 + \Delta g_1 + \Delta \Phi_1 \tag{106}$$

$$\dot{e}_2 = [A_2\big(\hat{x}_1, y\big) - S_2^{-1}C_2^T C_2]e_2 + \Delta A_2 x_2 + \Delta g_2 + \Delta \Phi_2 \tag{107}$$

where $\Delta A_1 = A_1(x_2, y) - A_1(\hat{x}_2, y)$,$\Delta g_1 = g_1(\Omega, y) - g_1(\hat{\Omega}, y)$, $\Delta \Phi_1 = \Phi_1(x_1, u) - \Phi_1(\hat{x}_1, u)$,$\Delta A_2 = A_2(x_1, y) - A_2\big(\hat{x}_1, y\big)$,$\Delta g_2 = g_2(\Omega, y) - g_2(\hat{\Omega}, y)$ and $\Delta \Phi_2 = \Phi_2(x_1, \Omega, u) - \Phi_2(\hat{x}_1, \hat{\Omega}, u)$.

Theorem 2 *Let's consider the PMSM dynamic model represented by Eq.* (99) *and* (100). *System* (101) *and* (102) *is an adaptive interconnected observer for system* (99) *and* (100) *with stability of estimation error dynamics. The speed adaptation law which guarantees the observer stability is derived as:*

$$\frac{d\hat{\Omega}}{dt} = \frac{1}{\lambda}[pi_d\left(e_{iq}S_{11}^{'} + e_{Rs}S_{12}^{'}\right) - pi_q(e_{id}S_{11} + e_{Ls^{-1}}S_{12})] \tag{108}$$

with $\lambda > 0$ and S_{11},S_{12},$S_{11}^{'}$,$S_{12}^{'}$ are the terms of the matrix S_1 and S_2 such as $S_1 = \begin{pmatrix} S_{11} & S_{12} \\ S_{12} & S_{22} \end{pmatrix}$ and $S_2 = \begin{pmatrix} S_{11}^{'} & S_{12}^{'} \\ S_{12}^{'} & S_{22}^{'} \end{pmatrix}$. $e_{Rs} = R_s - \hat{R}_s$ is the stator resistance estimation error and $e_{Ls} = L_s^{-1} - \hat{L}_s^{-1}$ is the inductance estimation error.

Proof To check convergence, let's consider the following Lyapunov equation:

$$V = V_1 + V_2 + \lambda(\Delta\Omega)^2 \tag{109}$$

where: $V_1 = e_1^T S_1 e_1$, $V_2 = e_2^T S_2 e_2$, $\Delta\Omega = \Omega - \hat{\Omega}$ and λ is a positive constant.

By calculating the derivative of V along trajectories of e_1 and e_2 we obtain:

$$\begin{aligned}
\dot{V} = {} & e_1^T[(A_1(\hat{x}_2, y) - S_1^{-1}C_1^T C_1)^T S_1 + S_1(A_1(\hat{x}_2, y) - S_1^{-1}C_1^T C_1) + \dot{S}_1]e_1 \\
& +2e_1^T S_1 \Delta A_1 x_1 + 2e_1^T S_1 \Delta\Phi_1 \\
& +e_2^T\left[(A_2(\hat{x}_1, y) - S_2^{-1}C_2^T C_2)^T S_2 + S_2(A_2(\hat{x}_1, y) - S_2^{-1}C_2^T C_2) + \dot{S}_2\right]e_2 \\
& +2e_2^T S_2 \Delta A_2 x_2 + 2e_2^T S_2 \Delta\Phi_2 \\
& +2e_1^T S_1 \Delta g_1 + 2e_2^T S_2 \Delta g_2 + 2\lambda\Delta\Omega\frac{d}{dt}\hat{\Omega}
\end{aligned} \tag{110}$$

One decomposes the function $\dot{V}$ in two functions σ_1 and σ_2 with σ_1 defined by:

$$\begin{aligned}
\sigma_1 = {} & e_1^T[(A_1(\hat{x}_2, y) - S_1^{-1}C_1^T C_1)^T S_1 + S_1(A_1(\hat{x}_2, y) - S_1^{-1}C_1^T C_1) + \dot{S}_1]e_1 \\
& +2e_1^T S_1 \Delta A_1 x_1 + 2e_1^T S_1 \Delta\Phi_1 \\
& +e_2^T\left[(A_2(\hat{x}_1, y) - S_2^{-1}C_2^T C_2)^T S_2 + S_2(A_2(\hat{x}_1, y) - S_2^{-1}C_2^T C_2) + \dot{S}_2\right]e_2 \\
& +2e_2^T S_2 \Delta A_2 x_2 + 2e_2^T S_2 \Delta\Phi_2
\end{aligned} \tag{111}$$

σ_2 isdefined by:

$$\sigma_2 = 2e_1^T S\Delta g_1 + 2e_2^T S_2 \Delta g_2 + 2\lambda\Delta\Omega\frac{d}{dt}\hat{\Omega} \tag{112}$$

The stability of the observer is proved if the derivative of the Lyapunov equation is negative. If we consider $\sigma_1 < 0$ and $\sigma_2 = 0$, the stability condition of the observer is proved.

Introducing standards for function σ_1, we can write:

$$\begin{aligned} &\sigma_1 \leq -\theta_1 e_1^T S_1 e_1 - \theta_2 e_2^T S_2 e_2 \\ &+2\|e_1\|\|S_1\|\|\Delta A_1\|\|x_1\| \\ &+2\|e_1\|\|S_1\|\|\Delta \Phi_1\| \\ &+2\|e_2\|\|S_2\|\|\Delta A_2\|\|x_2\| \\ &+2\|e_2\|\|S_2\|\|\Delta \Phi_2\| \end{aligned} \tag{113}$$

Knowing that g_1,g_2 and A_1 are globally Lipschitz functions, we can write:

$$\begin{cases} \|\Delta A_1\| \leq \rho_1\|e_2\| \\ \|\Delta A_2\| \leq \rho_2\|e_1\| \\ \|\Delta \Phi_1\| \leq \rho_3\|e_1\| \\ \|\Delta \Phi_2\| \leq \rho_4\|e_1\| \end{cases} \tag{114}$$

where $\rho_i, i \in \{1, \ldots\ldots, 4\}$ are positive constants. The way to comput $\rho_j, j \in \{1, \ldots\ldots, 4\}$ is presented in Annex).

According to Lemma 1 and taking the initial conditions of the PMSM drive and the observer in the physical operation domain D_3, the following inequalities hold:

$$\begin{cases} \|S_1\| < k_1 \\ \|S_2\| < k_2 \\ \|x_1\| < k_3 \\ \|x_2\| < k_4 \end{cases} \tag{115}$$

where $k_i, i \in \{1, \ldots\ldots, 4\}$ are positive constants. The way to comput $\rho_j, j \in \{1, \ldots\ldots, 4\}$ is presented in Annex.

Assume that the known inputs of each observer are regularly persistent. According to Lemma 1, there exist real numbers $\lambda_{min}(S_1) > 0$,$\lambda_{\max}(S_1) > 0$,$\lambda_{min}(S_2) > 0$,$\lambda_{\max}(S_2) > 0$, such that:0

$$\begin{cases} \lambda_{min}(S_1)\|e_1\|^2 \leq V_1 \leq \lambda_{max}(S_1)\|e_1\|^2 \\ \lambda_{min}(S_2)\|e_2\|^2 \leq V_2 \leq \lambda_{max}(S_2)\|e_2\|^2 \end{cases} \tag{116}$$

The condition on the function $\sigma_1\sigma_1$ σ_1 becomes:

$$\sigma_1 \leq -(\theta_1 - \tau)V_1 - \theta_2 V_2 + \delta\sqrt{V_1}\sqrt{V_2} \tag{117}$$

where $\tau = \frac{2k_1\rho_3}{\lambda_{\min}(S_1)}$ and $\delta = \frac{2k_1k_3\rho_1+2k_4k_2\rho_2+2k_2\rho_4}{\sqrt{\lambda_{min}(S_1)}\sqrt{\lambda_{min}(S_2)}}$.

Considering Eq. (46), the condition on the function σ_1 defined on (117) becomes:

$$\sigma_1 \leq -\left(\theta_1 - \tau - \frac{\delta}{2}\right)V_1 - \left(\theta_2 - \frac{\delta}{2}\right)V_2 \tag{118}$$

or

$$\sigma_1 \leq -\xi(V_1 + V_2) \leq -\xi V \tag{119}$$

with $\xi = \min(\theta_1 - \tau - \frac{\delta}{2}, \theta_2 - \frac{\delta}{2})$.

The condition on σ_1 is verified if we choose θ_1 and θ_2 such that $\xi > 0$.

Considering the validation of the first condition, to ensure the observer stability, we must admit that σ_2 is equal to zero , then we obtain:

$$\frac{d\hat{\Omega}}{dt} = \frac{1}{\lambda}\left[e_2^T S_2\begin{pmatrix} pi_d \\ 0 \end{pmatrix} - e_1^T S_1\begin{pmatrix} pi_q \\ 0 \end{pmatrix}\right] \tag{120}$$

The adaptation law $\hat{\Omega}$ is defined by:

$$\frac{d\hat{\Omega}}{dt} = \frac{1}{\lambda}[pi_d\left(e_{iq}S'_{11} + e_{Rs}S'_{12}\right) - pi_q(e_{id}S_{11} + e_{Ls^{-1}}S_{12})] \tag{121}$$

with S_{11},S_{12},S'_{11},S'_{12} are the terms of the matrix S_1 and S_2 such as $S_1 = \begin{pmatrix} S_{11} & S_{12} \\ S_{12} & S_{22} \end{pmatrix}$ and $S_2 = \begin{pmatrix} S'_{11} & S'_{12} \\ S'_{12} & S'_{22} \end{pmatrix}$.

Another formulation of Eq. (69), we have:

$$\frac{d\hat{\Omega}}{dt} = f_1(x, \hat{x}) + f_2(R_s, \hat{R}_s, L_s^{-1}, \hat{L}_s^{-1}) \tag{122}$$

where $f_1(x, \hat{x}) = \frac{1}{\lambda}\left[e_{iq}S'_{11}pi_d - e_{id}S_{11}pi_q\right]$ and $f_2(R_s, \hat{R}_s, L_s^{-1}, \hat{L}_s^{-1}) = -\frac{p}{\lambda}\left[e_{Rs}S'_{12}pi_d - e_{Ls^{-1}}S_{12}pi_q\right]$.

If Eq. (122) is checked, the derivative of V becomes:

$$\dot{V} \leq 0 \tag{123}$$

So, System (101) and (102) is an adaptive interconnected observer for system (99) and (100) with the stability of estimation error dynamics and Eq. (121) is a speed adapted law that guarantees the observer stability.

5.3 Simulation Results

To investigate the mathematical study and, hence, to illustrate the performance of the suggested observer, Table 1 gives the PMSM nominal parameters which are used in the simulation. The simulation results have been carried out using Matlab/Simulink Software. The developed study is applied and tested to a wind energy conversion system (WECS) using a PMSM. Characteristics of the associated turbine are given in Table 3 (Annex). The wind profile used in the simulation is presented in Fig. 3.

Figures 25, 27 and 29 have shown that estimated and measured variables have the same shape with similar amplitude. The estimated magnitudes correctly follow their actual states very well. Figures 26, 28 and 30 have presented that the observer converges after an average of 1 s. The stability of the estimation error dynamics is clearly shown in Figures 26b, 28b and 30b. It can be deduced that the observer gives a desirable result. The stability analysis of the suggested observer was so proved by simulation results.

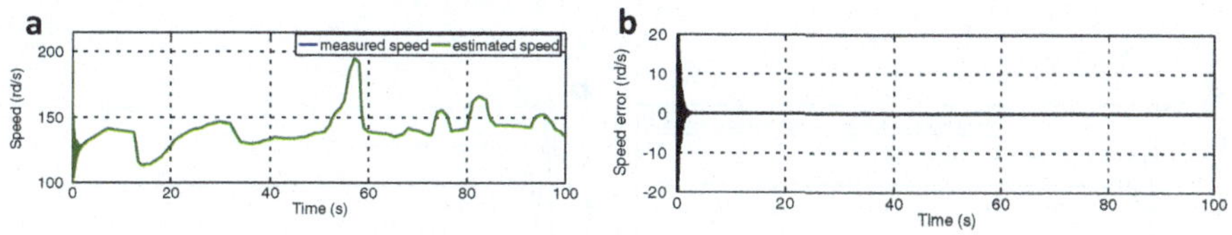

Fig. 25 Speed tracking: **a** real and estimated speed, **b**: speed error

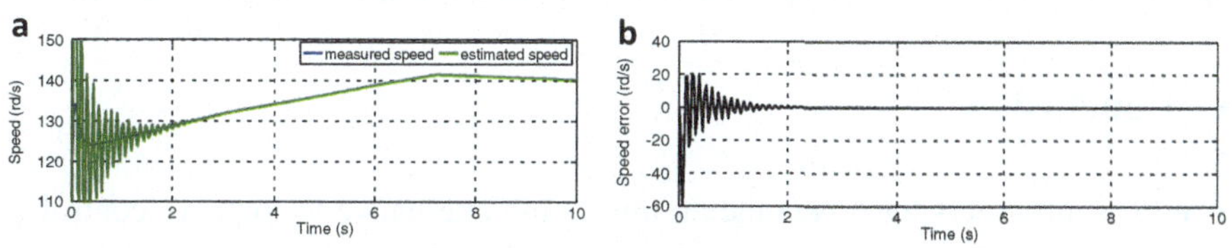

Fig. 26 Zoom of speed tracking: **a** real and estimated speed, **b**: speed error

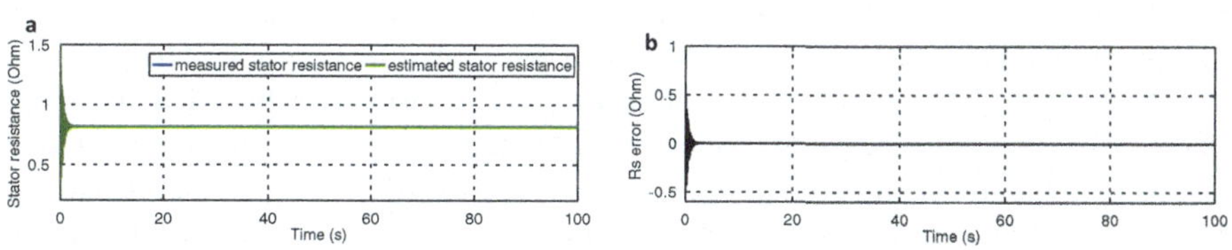

Fig. 27 Stator resistance estimation: **a** real and estimated stator resistance, **b**: stator resistance error

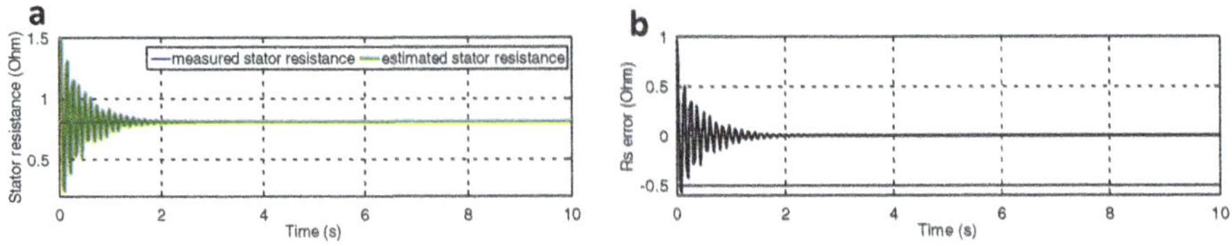

Fig. 28 Zoom of Stator resistance estimation: **a** real and estimated stator resistance, **b**: stator resistance error

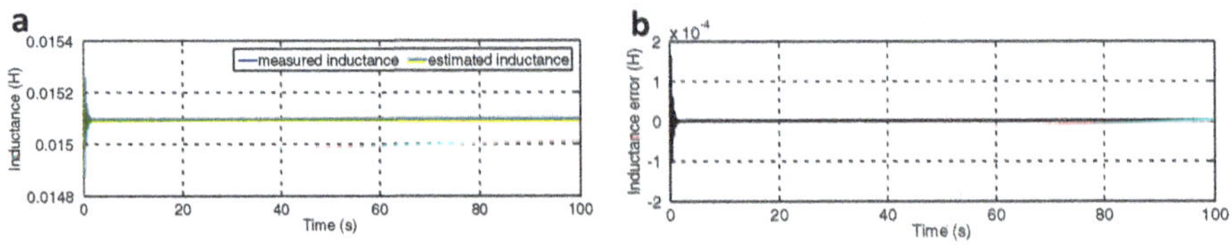

Fig. 29 Inductance estimation: **a**: real and estimated inductance, **b**: inductance error

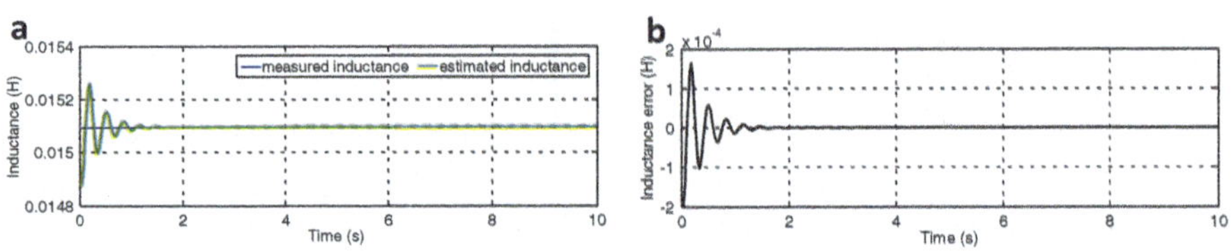

Fig. 30 Zoom of inductance estimation: **a**: real and estimated inductance, **b**: inductance error

6 Conclusion

In WECS, observers have become an interesting alternative to improve control of wind generators due to measurement errors that affect the system control quality and overall reliability. We have presented in these chapter three observations technique based on WECS namely: the adaptive observer, the interconnected observer and the adaptive interconnected observer. A comparative study is carried out between the suggested adaptive observer and the interconnected observer applied to IM based WECS. We have drawn that the interconnected observer is more robust than the adaptive one especially in the case of inductive parameters variation. On the other hand, it is to be noted that the value of stator resistance is required for stator flux estimation. Its variation due to frequency or temperature affects the scheme performance. To overcome this problem, a robust adaptive interconnected high gain observer with an online estimation of stator resistance applied to WECS using IM has been developed in the next part of this chapter. The stability and robustness study of the suggested observer have been described. According to simulation results, it can be deduced that stator resistance variation has been compensated by the new stator resistance estimator. Furthermore, it can be noted that, despite the parametric variation, the estimated magnitudes converge and give desirable results. The robustness of the adaptive high gain interconnected observer was thus validated. The final part of this chapter is concentrated on observer design for WECS based PMSM. The PMSM parameters are required for an eventual control. Its variation due to frequency or temperature affects the scheme performance. To overcome this problem, an adaptive interconnected high gain observer with online parametric estimation has been developed. The stability study of the suggested observer has been described. According to simulation results, it can be deduced that the observer converges and gives desirable results. The performance of the suggested adaptive interconnected observer for WECS based PMSM was thus validated.

References

1. GWEC Annual Report 2012 (2013) Global wind Energy Council. http://www.gwec.net/wp-content/uploads/2012/06/Annual_report_2012_LowRes.pdf. Accessed 25 July 2013
2. EWEA Report 2011 (2013) Pure power: wind energy targets for 2020 and 2030. The European Wind Energy Association 2011. http://www.ewea.org/fileadmin/ewea_documents/documents/publications/reports/Pure_Power_III.pdf. Accessed 25 July 2013
3. Kuchar, M., Brandstetter, P., Kaduch, M.: Sensorless induction motor drive with neural network PESC, pp. 3301–3305. Aachen Germany (2004)
4. Rong, W.J.: backstepping wavelet neural network control for indirect field-oriented induction motor drive. IEEE Trans. Neural Netw. **15**, 367–381 (2004)
5. Kwan, C.M., Lewis, F.L.: Robust backstepping control of induction motors using neural networks. IEEE Trans. Neural Netw. **11**, 1178–1187 (2000)
6. Morand, F.: Techniques d'observation sans capteur de vitesse en vue de la commande des machines asynchrones. Phdthesis in INSAL (2005)
7. Ferrah, A., Bradley, K.J., Hogben-Laing, P.J. et al.: Speed identifier for induction motor drives using real time adaptative digital filtering. IEEE Trans. Ind. Appl. **34**,156–162 (1998)
8. Holtz, J.: Is sensorless position control of standard induction motors a feasible technology. In: Proceeding of the Power Electronics and Motion Conference. IEEE, Beijing, China (2000)
9. Saheb, T.: Commande Sans Capteur Mécanique de la Machine Asynchrone. Phdthesis in IREENA (2004)
10. Besançon, G., Bornand, G., Hammouri, H.: Observer synthesis for a class of nonlinear control systems. Euro. J. Control **2**, 176–192 (1996)
11. Ghanes, M.: Observation et Commande de la Machine Asynchrone sans Capteur Mécanique. Phdthesis in IRCCyN (2005)
12. Ghanes, M., De Leon, J., Glumineau, A.: Observability Study and Observer-Based Interconnected from for Sensorless Induction Motor. CDC, San Diego, pp.1240–1244 (2006)
13. Schauder, C.: Adaptative speed identification for vector control of induction motors without rotational tranducers. IEEE Trans. Ind. Appl. **28**, 1054–1061 (1992)
14. Maes, J., Melkebeek, J.: Speed sensorless Direct Torque control of Induction Motor using adaptive flux observer. IEEE Trans. Ind. Appl. **36**, 778–7785 (2000)
15. Ghanes, M.: Observation et Commande de la Machine Asynchrone sans Capteur Mécanique, thèse de doctorat, IRCCyN (Ecole Centrale de Nantes) (2005)
16. Boukettaya, G.: A comparative study of three different sensorless vector control strategies for a flywheel energy storage system. Energy **35**, 132–139 (2010)
17. Traore, D.: Commande nonlineaire sans capteur de la machine asynchrone. Thèse de doctorat, Ecole Centrale de Nantes (2008)

18. Chiasson, J.: Non linear controllers for induction motors. In: IFAC Conference System Structure and Control. Nantes (1995)
19. Pillay, P., Krishnan, R.: Modeling, simulation, and analysis of permanent-magnet motor drives. ii. The brushless dc motor drive. IEEE Trans. Ind. Appl. **25**(2), 274–279 (1989). https://doi.org/10.1109/28.25542
20. Muljadi, E., Butterfield, C.P.: Pitch-controlled variable-speed wind turbine generation. IEEE Trans. Ind. Appl. **37**, 240–246 (2001)
21. Wang, P., Billinton, R.: Reliability benefit analysis of adding WTG to a distribution system. IEEE Trans. Energy Convers. **16**, 134–139 (2001)
22. Chen, Z., Tomita, M., Doki, S., Okuma, S.: An extended electromotive force model for sensorless control of interior permanent-magnet synchronous motors. IEEE Trans. Ind. Electron. **50**(2), 288–295 (2003). https://doi.org/10.1109/TIE.2003.809391
23. Marwali, M.N., Keyhani, A.: A comparative study of rotor flux based MRAS and back EMF based MRAS speed estimators for speed sensorless vector control of induction machines. In: Proceedings of the IEEE-IAS Annual Meeting, pp. 6–160 (1997)

On Observers-Based Controller Design for Induction Machine

Omar Naifar and Ghada Boukettaya

Abstract This chapter is devoted to the stability investigation of a controlled IM using both adaptive and interconnected observers. The introduced robust control is validated by using the Lyapunov approach. The global stability analysis of the closed-loop system is proven under parametric variations. Finally, simulations results are given to show the performance of the suggested observer's based controller design for the induction machine.

Keywords Robust control design · Adaptive observer based controller schema · Interconnected observer based controller schema

1 Introduction

Thanks to the technology developments and the recent advances in control theory, it is possible to implement new controllers to a large number of Alternative Current (AC) machines. These controllers are more robust to uncertainties, and more efficient under a wide range of operating conditions in very useful applications. One of the most attractive applications of electrical machines is for transport: vehicle traction that is currently in important development. The control of AC machines is a challenging problem that has attracted attention thanks to its several applications. For instance, the control problem of the induction motor has recently attracted attention due to its complexity: the induction motor is a nonlinear multi-variable system.

For economic reasons, for operating safety or to a degraded but functional solution to applications in case of failure of these sensors, the control without mechanical sensor requires the attention of many manufacturers. Therefore, it has become a centre of research interest in recent years. In this view, the main objective of this study

O. Naifar (✉)
Control and Energy Management Laboratory, University of Sfax, National School of Engineering of Sfax, Sfax, Tunisia

G. Boukettaya
Reaserch Laboratory of Renewable Energy and Electric System (LSEER) National School of Engineers of Sfax (ENIS), 3038 Sfax, Tunisia

O. Naifar and A. Ben Makhlouf (eds.), *Advances in Observer Design and Observation for Nonlinear Systems*, Studies in Systems, Decision and Control 410,
https://doi.org/10.1007/978-3-030-92731-8_9

is to synthesize nonlinear control laws without mechanical sensors for asynchronous machines. Initially, much attention is given to observers (soft sensor) to compensate for the absence of the mechanical sensor giving the speed information and the torque load using the unique measurement of currents. Then sophisticated control laws (non-linear control) are developed and associated with the observers to carry out the command without the mechanical sensor of the asynchronous machine.

Trouble for the sensorless speed control is evidence of the global stability of the closed-loop system (Control + Observer). In literature, little efforts have offered an inclusive demonstration of this method, except [1–4]. In fact, the work of [1] presents a global tracking control for speed-sensorless induction motors. Authors in [1] have developed a new second-order control for speed sensorless induction motor which guarantee the global stability of the closed-loop system (control + observer). Actually, they have designed an open-loop control which is extended to a closed-loop solution integrating states estimation. M. Feemster et al. have presented in [2] a sensorless control algorithm that achieves semi-global exponential rotor velocity tracking for the full-order nonlinear dynamic model of an induction motor actuating a mechanical subsystem.

Changes in his thesis [3] have described three observation strategies. The first one has named: cascade observer interconnected to an estimator. The second one has named: high gain observer interconnected to an estimator. The third one is named: high gain interconnected observer. Each of the above-cited observers is validated by the vector control and the sliding mode control. Rigorous mathematical stability analysis for the closed-loop system is presented. D. Traoré et al. in [4] have extended the results of [3]. In fact, the proposed sensorless control was the adaptive interconnected observer-based backstepping control. The architecture of the observer in [4] is like that used in [3]. While-that the proposed control is the backstepping control.

2 Control Based on the Lyapunov Approach

In this section, a novel control method for IM has been established. The main idea of this technic is to make stable the controlled closed loop system. This technique compared with traditional IM control uses a simple computation by the fact that a global asymptotic stability is obtained. The goal of this control method is based on a suitable choice of the voltage control values that warranty a global stability of the closed loop system. The IM mechanical variables flawlessly track their references through two PI controllers. The associate PI gains are computed by pole placement mode. Robustness study against parametric variation is established.

To fix the control voltage $U = (u_{sd}, u_{sq})^T$, a likely method is to create a (PI) controllers respectively for flux and speed, the regulators output provides the currents reference $I^* = (i^*_{sd}, i^*_{sq})^T$. The global stability analysis in the sense of Lyapunov lets us to define the control law. The Fig. 1 shows the overall structure of the novel control method based on the Lyapunov theory.

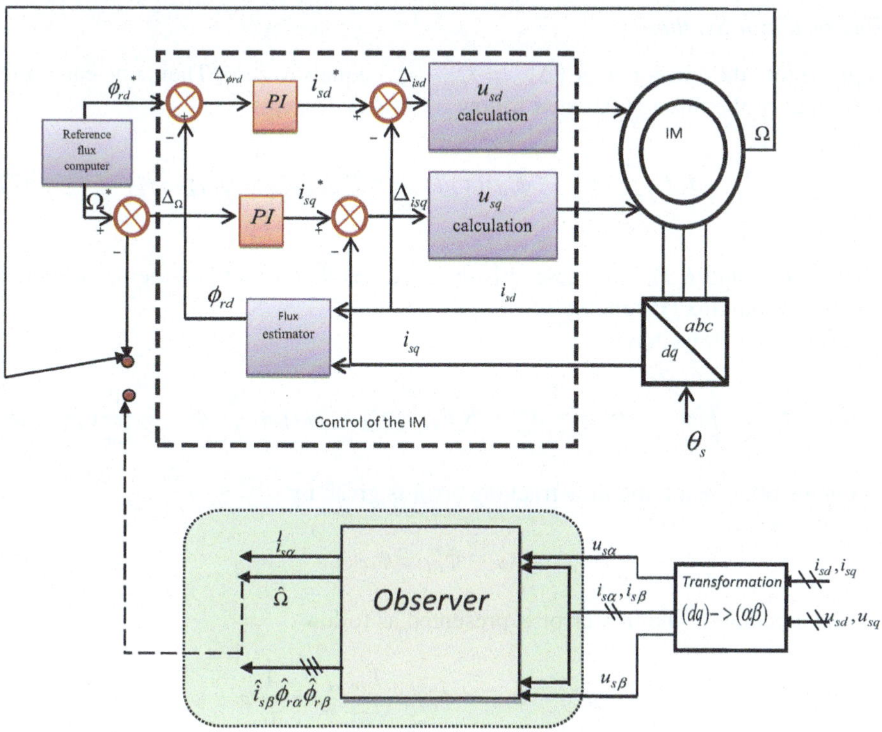

Fig. 1 General structure of the observer based controller for induction

We consider no variation on the induction machine parameters. We can name it the nominal case. For the stability analysis, we have defined below the Flux regulator synthesis and the speed regulator synthesis. A change of variable is made on the equations of both regulators to facilitate the determination of its gains and to easily analyze the stability in the sense of Lyapunov. A suitable choice of the control voltage confirms the asymptotic convergence of the system. To compensate nonlinearities, a possible method is to achieve a fast current loop (PI controller) [5], which will aim to force the current i_{sd} and i_{sq} to their reference values i_{sd}^* and i_{sq}^*. In this study, we suppose that the induction machine operates as a motor. For the control need, we use the induction machine model as it is presented in [5]. Then, the equations of the machine after the first loop are:

$$\begin{bmatrix} \dot{\Omega} \\ \dot{\phi}_{rd} \end{bmatrix} = \begin{bmatrix} m\phi_{rd} i_{sq}^* - \frac{T_l}{J} - c\Omega \\ \frac{L_m}{\tau_r} i_{sd}^* - \frac{1}{\tau_r}\phi_{rd} \end{bmatrix} \tag{1}$$

where $m = pL_m/JL_r$ and $c = f/J$.

Flux regulator Synthesis:

Suppose that the current error $\left(\Delta_{i_{sd}} = i_{sd}^* - i_{sd}\right)$ equals to zero. Then, one can write the output of the flux regulator as follow

$$i_{sd}^* = KI_{\phi rd} \int_0^t \left(\phi_{rd}^* - \phi_{rd}\right)(\tau)d\tau + KP_{\phi rd}\left(\phi_{rd}^* - \phi_{rd}\right) + i_{sd}^* \tag{2}$$

where $KI_{\phi rd}$ and $KP_{\phi rd}$ are respectively the integral constant and the proportional constant of the flux (PI) regulator.

Using Eq. (1), we have

$$i_{sd}^* = KI_{\phi rd} \int_0^t \left(\phi_{rd}^* - \phi_{rd}\right)(\tau)d\tau + KP_{\phi rd}\left(\phi_{rd}^* - \phi_{rd}\right) + \frac{\tau_r}{L_m}\dot{\phi}_{rd}^* + \frac{1}{L_m}\phi_{rd}^* \tag{3}$$

On the other hand, the flow tracking error is given by

$$\Delta_\phi = \phi_{rd}^* - \phi_{rd} \tag{4}$$

The dynamic of the flux error is presented as follow

$$\dot{\Delta}_\phi = \dot{\phi}_{rd}^* - \dot{\phi}_{rd} = \dot{\phi}_{rd}^* - \frac{L_m}{\tau_r} i_{sd}^* + \frac{1}{\tau_r}\phi_{rd} \tag{5}$$

Replacing i_{sd}^* by its expression, we have:

$$\dot{\Delta}_\phi = \frac{1}{\tau_r}\left(-1 - L_m KP_{\phi rd}\right)\Delta_\phi - \frac{L_m KI_{\phi rd}}{\tau_r} \int_0^t \Delta_\phi(\tau)d\tau \tag{6}$$

Consider the following coordinates change:

$$\Gamma_\phi = \left(\int_0^t \Delta_\phi(\tau)d\tau, \Delta_\phi\right)^T \tag{7}$$

The dynamics of the flux error (5) in the new coordinates is given by:

$$\dot{\Gamma}_\phi = A_\phi \Gamma_\phi \tag{8}$$

where $A_\phi = \begin{pmatrix} 0 & 1 \\ \beta_{1\phi} & \beta_{2\phi} \end{pmatrix}$ with $\beta_{1\phi} = -\frac{L_m KI_{\phi rd}}{\tau_r}$ and $\beta_{2\phi} = \frac{1}{\tau_r}\left(-1 - L_m KP_{\phi rd}\right)$.

The gains $KI_{\phi rd}$ and $KP_{\phi rd}$ are determined in such ways that the matrix A_ϕ is stable. See Annex.

Speed regulator Synthesis:

Once the machine is fluxed $\left(\phi_{rd} = \phi_{rd}^{*} = constant\right)$, the equation of the electromechanical torque become $C_{em} = K_T i_{sq}^{*}$ with K_T is an electromagnetic torque constant defined by $p\frac{L_m}{L_r}\phi_{rd}$. Suppose that the current error $\left(\Delta_{i_{sq}} = i_{sq}^{*} - i_{sq}\right)$ equals to zero. Then, one can write the output of the flux regulator as follow:

$$
\begin{aligned}
i_{sq}^{*} &= \frac{1}{K_T}\left[K I_{\Omega} \int_0^t \left(\Omega^{*} - \Omega\right)(\tau)d\tau + K P_{\Omega}\left(\Omega^{*} - \Omega\right)\right] + i_{sq}^{*} \\
i_{sq}^{*} &= \frac{1}{K_T}\left[K I_{\Omega} \int_0^t \left(\Omega^{*} - \Omega\right)(\tau)d\tau + K P_{\Omega}\left(\Omega^{*} - \Omega\right)\right] + \frac{1}{m\phi_{rd}}\left[\dot{\Omega}^{*} + c\Omega + \frac{T_l}{J}\right]
\end{aligned}
\tag{9}
$$

where $K I_{\Omega}$ and $K P_{\Omega}$ are respectively the integral constant and the proportional constant of the speed (PI) regulator.

On the other hand, the speed tracking error is given by

$$
\Delta_{\Omega} = \Omega - \Omega^{*} \tag{10}
$$

The dynamic of the flux error is presented as follow

$$
\dot{\Delta}_{\Omega} = \dot{\Omega} - \dot{\Omega}^{*} = m\phi_{rd} i_{sq}^{*} - \frac{T_l}{J} - c\Omega - \dot{\Omega}^{*} \tag{11}
$$

Replacing i_{sq}^{*} by its expression, we have:

$$
\dot{\Delta}_{\Omega} = -\frac{K P_{\Omega}}{J}\Delta_{\Omega} - \frac{K I_{\Omega}}{J}\int_0^t \Delta_{\Omega}(\tau)d\tau \tag{12}
$$

Consider the following coordinates change:

$$
\Gamma_{\Omega} = \left(\int_0^t \Delta_{\Omega}(\tau)d\tau,\ \Delta_{\Omega}\right)^T \tag{13}
$$

The dynamics of the speed error (11) in the new coordinates is given by:

$$
\dot{\Gamma}_{\Omega} = A_{\Omega}\Gamma_{\Omega} \tag{14}
$$

where $A_{\Omega} = \begin{pmatrix} 0 & 1 \\ \beta_{1\Omega} & \beta_{2\Omega} \end{pmatrix}$ with $\beta_{1\Omega} = -\frac{K I_{\Omega}}{J}$ and $\beta_{2\Omega} = -\frac{K P_{\Omega}}{J}$.

The gains KI_Ω and KP_Ω are determined in such ways that the matrix A_Ω is stable. See Annex.

The dynamic currents errors are defined by:

$$\dot{\Delta}_{i_{sd}} = \dot{i}_{sd}^* - \dot{i}_{sd} = \dot{i}_{sd}^* - \delta_1 - \frac{1}{\sigma L_s} u_{sd} \tag{15}$$

$$\dot{\Delta}_{i_{sq}} = \dot{i}_{sq}^* - \dot{i}_{sq} = \dot{i}_{sq}^* - \delta_2 - \frac{1}{\sigma L_s} u_{sq} \tag{16}$$

where: $\Delta_{i_{sd}} = i_{sd}^* - i_{sd}$; $\Delta_{i_{sq}} = i_{sq}^* - i_{sq}$; $\delta_1 = -\gamma_1 i_{sd} + \frac{\gamma_2}{\tau_r}\phi_{rd} + p\Omega i_{sq} + \frac{L_m}{\tau_r}\frac{i_{sq}^2}{\phi_{rd}}$; and $\delta_2 = -\gamma_1 i_{sq} - p\Omega i_{sd} - p\Omega\gamma_2\phi_{rd} - \frac{L_m}{\tau_r}\frac{i_{sd} i_{sq}}{\phi_{rd}}$.

2.1 Stability Study in Case of Rated Operation

Since A_ϕ and A_Ω are two stable matrixes, so $\forall Q_\phi > 0$ and $Q_\Omega > 0\ \exists P_\phi = P_\phi^T > 0$ and $P_\Omega = P_\Omega^T > 0$ defined as:

$$P_\phi A_\phi + A_\phi^T P_\phi = -Q_\phi \ \ and \ \ P_\Omega A_\Omega + A_\Omega^T P_\Omega = -Q_\Omega \tag{17}$$

To check the convergence of the proposed control technique, we check the convergence of different dynamics associated respectively with flux, speed and currents errors. So we consider the following Lyapunov function:

$$V_c = \Gamma_\phi^T P_\phi \Gamma_\phi + \Gamma_\Omega^T P_\Omega \Gamma_\Omega + \frac{1}{2}(\Delta_{i_{sd}}^2 + \Delta_{i_{sq}}^2) \tag{18}$$

By calculating the derivative of V_c we obtain:

$$\dot{V}_c = \Gamma_\phi^T (P_\phi A_\phi + A_\phi^T P_\phi)\Gamma_\phi + \Gamma_\Omega^T (P_\Omega A_\Omega + A_\Omega^T P_\Omega)\Gamma_\Omega + \Delta_{i_{sd}}\dot{\Delta}_{i_{sd}} + \Delta_{i_{sq}}\dot{\Delta}_{i_{sq}} \tag{19}$$

Equation (19) becomes:

$$\dot{V}_c = -\Gamma_\phi^T Q_\phi \Gamma_\phi - \Gamma_\Omega^T Q_\Omega \Gamma_\Omega + \Delta_{i_{sd}}\left(K_{i_{sd}}\Delta_{i_{sd}} + \dot{i}_{sd}^* - \delta_1 - \frac{1}{\sigma L_s}u_{sd}\right)$$
$$+ \Delta_{i_{sq}}\left(K_{i_{sq}}\Delta_{i_{sq}} + \dot{i}_{sq}^* - \delta_2 - \frac{1}{\sigma L_s}u_{sq}\right) - K_{isd}\Delta_{i_{sd}}^2 - K_{isq}\Delta_{i_{sq}}^2 \tag{20}$$

where $K_{i_{sd}}$, $K_{i_{sq}}$ are two positive constants.

Now, considering that inequality (21), (22), (23) and (24) are verified:

$$\lambda_{min}(P_\phi)\left\|\Gamma_\phi\right\|^2 \leq \Gamma_\phi^T P_\phi \Gamma_\phi \leq \lambda_{max}(P_\phi)\left\|\Gamma_\phi\right\|^2 \tag{21}$$

$$\lambda_{min}(Q_\phi)\left\|\Gamma_\phi\right\|^2 \leq \Gamma_\phi^T Q_\phi \Gamma_\phi \leq \lambda_{max}(Q_\phi)\left\|\Gamma_\phi\right\|^2 \tag{22}$$

$$\lambda_{min}(P_\Omega)\|\Gamma_\Omega\|^2 \leq \Gamma_\Omega^T P_\Omega \Gamma_\Omega \leq \lambda_{max}(P_\Omega)\|\Gamma_\Omega\|^2 \tag{23}$$

$$\lambda_{min}(Q_\Omega)\|\Gamma_\Omega\|^2 \leq \Gamma_\Omega^T Q_\Omega \Gamma_\Omega \leq \lambda_{max}(Q_\Omega)\|\Gamma_\Omega\|^2 \tag{24}$$

According to Eq. (20), the control voltages u_{sd} and u_{sq} are chosen as:

$$u_{sd} = \sigma L_s\left(K_{i_{sd}}\Delta_{i_{sd}} + \dot{i}_{sd}^* - \delta_1\right) \tag{25}$$

$$u_{sq} = \sigma L_s\left(K_{i_{sq}}\Delta_{i_{sq}} + \dot{i}_{sq}^* - \delta_2\right) \tag{26}$$

So, considering boundaries defined from (21) to (24) and using the fixed control voltages defined in (25) and (26), the inequality of the Lyapunov function derivative of the system (20) becomes:

$$\dot{V}_c \leq -\eta_\phi \Gamma_\phi^T P_\phi \Gamma_\phi - \eta_\Omega \Gamma_\Omega^T P_\Omega \Gamma_\Omega - 2K_{i_{sd}}\left(\frac{1}{2}\Delta_{i_{sd}}^2\right) - 2K_{i_{sq}}\left(\frac{1}{2}\Delta_{i_{sq}}^2\right) \tag{27}$$

where $\eta_\phi = \frac{\lambda_{min}(Q_\phi)}{\lambda_{max}(P_\phi)}$ and $\eta_\Omega = \frac{\lambda_{min}(Q_\Omega)}{\lambda_{max}(P_\Omega)}$.

We set $\theta = min(\eta_\phi, \eta_\Omega, 2K_{i_{sd}}, 2K_{i_{sq}})$, we obtain:

$$\dot{V}_c \leq -\theta V_c \tag{29}$$

The errors converge exponentially. So the choice of the control $U = (u_{sd}, u_{sq})^T$ confirms the stability of the controlled closed loop system.

2.2 Stability Study in Case of Uncertain Parameters

In this section, we consider a variation in the induction machine parameters. Generally, the rated operation is an ideal case and to confirm the robustness of the proposed control strategies, it must be proved and tested in the case of uncertain parameters. For the stability analysis, we have defined the uncertainty terms on both flux and speed regulators. Some conditions are presented with the same control law which leads to the asymptotic convergence of the system.

The dynamic of the flux, speed and currents error under uncertain parameters became:

$$\dot{\Gamma}_\phi = (A_\phi + \delta_\phi)\Gamma_\phi \tag{30}$$

$$\dot{\Gamma}_\Omega = (A_\Omega + \delta_\Omega)\Gamma_\Omega \tag{31}$$

$$\dot{\Delta}_{i_{sd}} = \dot{i_{sd}}^* - \dot{i}_{sd} = \dot{i}_{sd}^* - \delta_1 - \Theta\delta_1 - \frac{1}{\sigma L_s} u_{sd} - \Theta U\ u_{sd} \tag{32}$$

$$\dot{\Delta}_{i_{sq}} = \dot{i_{sq}}^* - \dot{i}_{sq} = \dot{i}_{sq}^* - \delta_2 - \Theta\delta_2 - \frac{1}{\sigma L_s} u_{sq} - \Theta U\ u_{sq} \tag{33}$$

where δ_ϕ, δ_Ω, $\Theta\delta_1$, $\Theta\delta_2$ and ΘU are respectively the uncertain term associated to A_ϕ, A_Ω, δ_1, δ_2 and $\frac{1}{\sigma L_s}$.The uncertain terms $|\delta_\phi|$, $|\delta_\Omega|$,$|\Theta\delta_1|$, $|\Theta\delta_2|$ and $|\Theta U|$ are assumed to be bounded respectively by the positive constant α_1, α_2, v_1, v_2 and $\tilde{n}$.

Including standards, Eq. (27) becomes:

$$\begin{aligned} \dot{V}_c \le & -\Gamma_\phi^T Q_\phi \Gamma_\phi - \Gamma_\Omega^T Q_\Omega \Gamma_\Omega + 2\|\Gamma_\phi\| \|P_\phi\| |\delta_\phi| + 2\|\Gamma_\Omega\| \|P_\Omega\| |\delta_\Omega| \\ & + \Delta_{i_{sd}} \left(K_{i_{sd}} \Delta_{i_{sd}} + \dot{i}_{sd}^* - \delta_1 - \Theta\delta_1 - \frac{1}{\sigma L_s} u_{sd} - \Theta U\ u_{sd} \right) \\ & + \Delta_{i_{sq}} \left(K_{i_{sq}} \Delta_{i_{sq}} + \dot{i}_{sq}^* - \delta_2 - \Theta\delta_2 - \frac{1}{\sigma L_s} u_{sq} - \Theta U\ u_{sq} \right) - K_{i_{sd}} \Delta_{i_{sd}}^2 - K_{i_{sq}} \Delta_{i_{sq}}^2 \end{aligned} \tag{34}$$

One notes that $h_1 = \Gamma_\phi^T P_\phi \Gamma_\phi$, $h_2 = \Gamma_\Omega^T P_\Omega \Gamma_\Omega$ and $h = h_1 + h_2$, one chooses the control voltage define in (25) and (26):

$$\begin{aligned} \dot{V}_c \le & -\eta_\phi h_1 - \eta_\Omega h_2 + 2\alpha_1 \lambda_{max}(P_\phi) \|\Gamma_\phi\| + 2\alpha_2 \lambda_{max}(P_\Omega) \|\Gamma_\Omega\| \\ & + \|\Delta_{isd}\| (-\Theta\delta_1 - \Theta U u_{sd}) + \|\Delta_{i_{sq}}\| (-\Theta\delta_2 - \Theta U u_{sq}) \\ & - K_{i_{sd}} \Delta_{isd}^2 - K_{i_{sq}} \Delta_{i_{sq}}^2 \end{aligned} \tag{35}$$

or

$$\begin{aligned} \dot{V}_c \le & -\eta_\phi h_1 - \eta_\Omega h_2 + \frac{2\alpha_1 \lambda_{max}(P_\phi)}{\sqrt{\lambda_{min}(P_\phi)}} \sqrt{h_1} + \frac{2\alpha_2 \lambda_{max}(P_\Omega)}{\sqrt{\lambda_{min}(P_\Omega)}} \sqrt{h_2} + \|\Delta_{i_{sd}}\| (-\Theta\delta_1 - \Theta U u_{sd}) \\ & + \|\Delta_{i_{sq}}\| (-\Theta\delta_2 - \Theta U u_{sq}) - K_{i_{sd}} \Delta_{i_{sd}}^2 - K_{i_{sq}} \Delta_{i_{sq}}^2 \end{aligned} \tag{36}$$

One notes that $\varsigma = \min(\eta_\phi, \eta_\Omega)$, $\mu = \max\left(\frac{2\alpha_1 \lambda_{max}(P_\phi)}{\sqrt{\lambda_{min}(P_\phi)}}, \frac{2\alpha_2 \lambda_{max}(P_\Omega)}{\sqrt{\lambda_{min}(P_\Omega)}} \right)$, inequality (36) become:

$$\dot{V}_c \leq -\varsigma h + \mu\gamma\sqrt{h} + \left\|\Delta_{i_{sd}}\right\|(-\Theta\delta_1 - \Theta U u_{sd}) + \left\|\Delta_{i_{sq}}\right\|\left(-\Theta\delta_2 - \Theta U u_{sq}\right) - K_{i_{sd}}\Delta_{i_{sd}}^2 - K_{i_{sq}}\Delta_{i_{sq}}^2 \tag{37}$$

γ is defined as: $\sqrt{h_1} + \sqrt{h_2} \leq \gamma\sqrt{h}$.

We can write another formulation of Eq. (37) defined by:

$$\dot{V}_c \leq -(1-\Upsilon)\varsigma h + \Delta_{i_{sd}}(-\Theta\delta_1 - \Theta U\ u_{sd}) + \Delta_{i_{sq}}\left(-\Theta\delta_2 - \Theta U\ u_{sq}\right) - K_{i_{sd}}\Delta_{i_{sd}}^2 - K_{i_{sq}}\Delta_{i_{sq}}^2 \qquad \forall\Gamma \geq \frac{\mu\gamma}{\Upsilon\varsigma};\ 0 < \Upsilon < 1 \tag{38}$$

where $\Gamma = \Gamma_\phi + \Gamma_\Omega$ is the vector of flux and speed errors.

By choosing $\varrho u_{sdmax} + \upsilon_1 > 0$ and $\varrho u_{sqmax} + \upsilon_2 > 0$, the Lyapunov function derivative $\dot{V}_c$ become negative. So the stability of the controlled closed loop system is verified.

3 Adaptive Observer–Controller Scheme Stability Analysis

The main goal is to achieve control without a mechanical sensor for the IM. Speed and flux are not measured, so the outputs of the speed and flux regulator will be presented as follow:

$$i_{sq}^*\left(\hat{\Omega}, \hat{\phi}_{rd}\right) = \frac{1}{K_T}\left[K I_\Omega \int_0^t \left(\Omega^* - \hat{\Omega}\right)(\tau)d\tau + K P_\Omega\left(\Omega^* - \hat{\Omega}\right)\right] + \frac{1}{m\hat{\phi}_{rd}}\left[\dot{\Omega}^* + c\hat{\Omega} + \frac{\widehat{T}_l}{J}\right] \tag{39}$$

$$i_{sd}^*\left(\hat{\phi}_{rd}\right) = K I_{\phi rd}\int_0^t\left(\phi_{rd}^* - \hat{\phi}_{rd}\right)(\tau)d\tau + K P_{\phi rd}\left(\phi_{rd}^* - \hat{\phi}_{rd}\right) + \frac{\tau_r}{L_m}\dot{\phi}_{rd}^* + \frac{1}{L_m}\phi_{rd}^* \tag{40}$$

where $\widehat{\Omega}$ and $\hat{\phi}_{rd}$ are the estimated value of the speed and flux given by the adaptive observer.

The dynamic tracking errors flux (8) and speed (13) become:

$$\begin{cases} \dot{\Gamma}_\phi = A_\phi\Gamma_\phi + B_\phi\chi_\phi(\varepsilon_\phi) \\ \dot{\Gamma}_\Omega = A_\Omega\Gamma_\Omega + B_\Omega\chi_\Omega(\varepsilon_\Omega) \end{cases} \tag{41}$$

where

$$\varepsilon_\phi = \phi_{rd} - \hat{\phi}_{rd},$$

$$\chi_\phi(\varepsilon_\phi) = -\frac{L_m}{\tau_r}\left[KP_{\phi rd}\varepsilon_\phi + KI_{\phi rd}\int_0^t \varepsilon_\phi(\tau)d\tau\right], \chi_\Omega(\varepsilon_\Omega)$$
$$= \chi_1(\varepsilon_\Omega) + \chi_2(\varepsilon_\phi) + \chi_3(\varepsilon_\Omega, \varepsilon_\phi),$$

$$\chi_1(\varepsilon_\Omega) = \frac{KI_\Omega}{J}\int_0^t \varepsilon_\Omega(\tau)d\tau + \left[\frac{KP_\Omega}{J} - c\right]\varepsilon_\Omega,$$

$$\chi_2(\varepsilon_\phi) = \frac{\varepsilon_\phi}{\hat{\phi}_{rd}}\left[\Omega^* + c\hat{\Omega} + \frac{\hat{T}_l}{J}\right] - \frac{KP_\Omega}{J}\left(\frac{\varepsilon_\phi}{\hat{\phi}_{rd}}\right)\Delta_\Omega - \frac{KI_\Omega}{J}\left(\frac{\varepsilon_\phi}{\hat{\phi}_{rd}}\right)\int_0^t \Delta_\Omega(\tau)d\tau,$$

$$B_\phi = B_\Omega = \begin{bmatrix} 0 \\ 1 \end{bmatrix},$$

$$\chi_3(\varepsilon_\Omega, \varepsilon_\phi) = \frac{\varepsilon_\phi}{\hat{\phi}_{rd}}\left[\frac{KP_\Omega}{J}\varepsilon_\Omega + \frac{KI_\Omega}{J}\int_0^t \varepsilon_\Omega(\tau)d\tau\right],$$

Theorem 1 Consider the induction machine model as in [5]. If the speed and flux regulators use the estimated variables given by the adaptive observer defined as in [5], then the errors of flux and speed asymptotically converge to zero.

Proof The control voltage u_{sd} and u_{sq} in this case are chosen as:

$$u_{sd} = \sigma L_s\left(K_{isd}\varepsilon_{isd} + \dot{i}^*_{sd}\left(\hat{\phi}_{rd}\right) - \delta_1\right) \tag{42}$$

$$u_{sq} = \sigma L_s\left(K_{isq}\varepsilon_{isq} + \dot{i}^*_{sq}\left(\hat{\Omega}, \hat{\phi}_{rd}\right) - \delta_2\right) \tag{43}$$

with $\varepsilon_{isd} = i^*_{sd}\left(\hat{\phi}_{rd}\right) - i_{sd}$; $\varepsilon_{isq} = i^*_{sq}\left(\hat{\Omega}, \hat{\phi}_{rd}\right) - i_{sq}$.

Consider the following Lyapunov function:

$$V_{ac} = V_a + V_c = \Upsilon + \lambda(\Delta\Omega)^2 + \Gamma_\phi^T P_\phi \Gamma_\phi + \Gamma_\Omega^T P_\Omega \Gamma_\Omega + \frac{1}{2}(\varepsilon_{isd}^2 + \varepsilon_{isq}^2) \tag{44}$$

Admitting the equation of the adaptation law as in [4]. From inequality (14) in [6], we have

$$\dot{V}_a = \sigma_1 < -\delta_0\|e\|^2; \forall\|e\| \geq \frac{\eta_2}{\varepsilon\eta_1\sqrt{\lambda_{min}(P)}}$$

with: $\delta_0 = -(1-\varepsilon)\eta_1$.

The derivative of V_{ac} is given by:

$$\dot{V}_{ac} \leq -\delta_0\|e\|^2 + \Gamma_\phi^T(P_\phi A_\phi + A_\phi^T P_\phi)\Gamma_\phi + \Gamma_\Omega^T(P_\Omega A_\Omega + A_\Omega^T P_\Omega)\Gamma_\Omega$$
$$+2\Gamma_\phi^T P_\phi B_\phi \chi_\phi(e_\phi) + 2\Gamma_\Omega^T P_\Omega B_\Omega \chi_\Omega(e_\Omega) + \varepsilon_{isd}\left(K_{isd}\varepsilon_{isd} + \dot{i}_{sd}^* - \delta_1 - \frac{1}{\sigma L_s}u_{sd}\right)$$
$$+\varepsilon_{isq}\left(K_{isq}\varepsilon_{isq} + \dot{i}_{sq}^* - \delta_2 - \frac{1}{\sigma L_s}u_{sq}\right) - K_{isd}\varepsilon_{isd}^2 - K_{isq}\varepsilon_{isq}^2 \tag{45}$$

Replacing the control voltages by its value, Eq. (45) became:

$$\dot{V}_{ac} \leq -\delta_0\|e\|^2 - \eta_\phi\left\|\Gamma_\phi\right\|^2 - \eta_\Omega\|\Gamma_\Omega\|^2 + 2l_1\left\|\Gamma_\phi\right\|\|e\| + 2l_2\|\Gamma_\Omega\|\|e\|$$
$$- K_{isd}\varepsilon_{isd}^2 - K_{isq}\varepsilon_{isq}^2 \tag{46}$$

where $\left\|\chi_\phi(\varepsilon_\phi)\right\| \leq l_1\|e\|$ and $\|\chi_\Omega(\varepsilon_\Omega)\| \leq l_2\|e\|$; $l_1 > 0$, $l_2 > 0$.

Considering the following inequalities:

$$\|e\|\left\|\Gamma_\phi\right\| \leq \frac{\zeta_1}{2}\left\|\Gamma_\phi\right\|^2 + \frac{1}{2\zeta_1}\|e\|^2;\ \|e\|\|\Gamma_\Omega\| \leq \frac{\zeta_2}{2}\|\Gamma_\Omega\|^2 + \frac{1}{2\zeta_2}\|e\|^2 \forall \zeta_1, \zeta_2 \in]0,1[\tag{47}$$

We obtains:

$$\dot{V}_{ac} \leq -\delta_0\|e\|^2 - \eta_\phi\left\|\Gamma_\phi\right\|^2 - \eta_\Omega\|\Gamma_\Omega\|^2 + l_1\zeta_1\left\|\Gamma_\phi\right\|^2 + l_2\zeta_2\|\Gamma_\Omega\|^2$$
$$- K_{isd}\varepsilon_{isd}^2 - K_{isq}\varepsilon_{isq}^2 + \frac{l_1}{\zeta_1}\|e\|^2 + \frac{l_2}{\zeta_2}\|e\|^2 \tag{48}$$

Grouping the various terms, we obtain:

$$\dot{V}_{ac} \leq -\left(\delta_0 - \frac{l_1}{\zeta_1} - \frac{l_2}{\zeta_2}\right)\|e\|^2 - (\eta_\phi - l_1\zeta_1)\left\|\Gamma_\phi\right\|^2 - (\eta_\Omega - l_2\zeta_2)\|\Gamma_\Omega\|^2$$
$$- 2K_{isd}\left(\frac{1}{2}\varepsilon_{isd}^2\right) - 2K_{isq}\left(\frac{1}{2}\varepsilon_{isq}^2\right) \tag{49}$$

We define $\upsilon = \min(\sigma_1, \sigma_2, \sigma_3, 2K_{isq}, 2K_{isd})$ where $\sigma_1 = \delta_0 - \frac{l_1}{\zeta_1} - \frac{l_2}{\zeta_2}$, $\sigma_2 = \eta_\phi - l_1\zeta_1$, $\sigma_3 = \eta_\Omega - l_2\zeta_2$.

The derivative of V_{oc} becomes:

$$\dot{V}_{oc} \leq -\upsilon V_{oc} \tag{50}$$

By choosing η_ϕ, η_Ω and δ_0 such that σ_1, σ_2 and σ_3 are greater than zero. Then, flux and speed errors converge asymptotically to zero.

4 Interconnected Observer–Controller Scheme Stability Analysis

In this part, we have proposed the interconnected observer for the proof of the global stability of the output feedback system.

The dynamic error of the closed loop system including the interconnected observer is given by:

$$\begin{cases} \dot{\Gamma}_\phi = A_\phi \Gamma_\phi + B_\phi \chi_\phi(\varepsilon_\phi) \\ \dot{\Gamma}_\Omega = A_\Omega \Gamma_\Omega + B_\Omega \chi_\Omega(\varepsilon_\Omega) \\ \dot{e}_1 = [A_1(\hat{x}_2) - \Gamma^{-1}(\hat{x}_2) S_1^{-1} C_1^T C_1] e_1 + g_1(u, x_2, x_1) - g_1(u, \hat{x}_2, \hat{x}_1) + \delta g_1(u, x_2, x_1) \\ +[A_1(x_2) - A_1(\hat{x}_2) + \delta A_1(x_2)] x_1 \\ \dot{e}_2 = [A_2(x_1) + \delta A_2(x_1)] x_2 - A_2(\hat{x}_1)\hat{x}_2 - S_2^{-1}(\theta_2, \hat{x}_1) C_2^T C_2 e_2 \end{cases} \tag{51}$$

Theorem 2 Consider the induction machine model (1) in [6]. If the speed and flux regulators use the estimated variables given by the interconnected observer (18)–(19) in [6], then the errors of flux and speed asymptotically converge to zero.

Proof The control voltage u_{sd} and u_{sq} in this case are chosen as:

$$u_{sd} = \sigma L_s \left(K_{isd} \varepsilon_{isd} + \dot{i}_{sd}^* \left(\hat{\phi}_{rd} \right) - \delta_1 \right) \tag{52}$$

$$u_{sq} = \sigma L_s \left(K_{isq} \varepsilon_{isq} + \dot{i}_{sq}^* \left(\widehat{\Omega}, \hat{\phi}_{rd} \right) - \delta_2 \right) \tag{53}$$

with $\varepsilon_{isd} = i_{sd}^* \left(\hat{\phi}_{rd} \right) - i_{sd}$; $\varepsilon_{isq} = i_{sq}^* \left(\widehat{\Omega}, \hat{\phi}_{rd} \right) - i_{sq}$.

Consider the following Lyapunov function:

$$V_{ic} = V_i + V_c = V_1 + V_2 + \Gamma_\phi^T P_\phi \Gamma_\phi + \Gamma_\Omega^T P_\Omega \Gamma_\Omega + \frac{1}{2}(\varepsilon_{isd}^2 + \varepsilon_{isq}^2) \tag{54}$$

Admitting the Eq. (39) in [4], we have $\dot{V}_i \leq -(1-\varepsilon)\lambda_1 V \leq -\delta_0 V$; $\forall \|e\| \geq \frac{\lambda_2' \psi}{\varepsilon \lambda_1'}$ with: $\delta_0 = (1-\varepsilon)\lambda_1$.

The derivative of V_{ic} is given by:

$$\begin{aligned} \dot{V}_{ic} \leq & -\delta_0 \|e\|^2 + \Gamma_\phi^T (P_\phi A_\phi + A_\phi^T P_\phi) \Gamma_\phi + \Gamma_\Omega^T (P_\Omega A_\Omega + A_\Omega^T P_\Omega) \Gamma_\Omega \\ & + 2\Gamma_\phi^T P_\phi B_\phi \chi_\phi(e_\phi) + 2\Gamma_\Omega^T P_\Omega B_\Omega \chi_\Omega(e_\Omega) + \varepsilon_{isd} \left(K_{isd} \varepsilon_{isd} + \dot{i}_{sd}^* - \delta_1 - \tfrac{1}{\sigma L_s} u_{sd} \right) \\ & + \varepsilon_{isq} \left(K_{isq} \varepsilon_{isq} + \dot{i}_{sq}^* - \delta_2 - \tfrac{1}{\sigma L_s} u_{sq} \right) - K_{isd} \varepsilon_{isd}^2 - K_{isq} \varepsilon_{isq}^2 \end{aligned} \tag{55}$$

Replacing the control voltages by its value, Eq. (55) became:

$$\dot{V}_{ic} \leq -\delta_0\|e\|^2 - \eta_\phi\left\|\Gamma_\phi\right\|^2 - \eta_\Omega\|\Gamma_\Omega\|^2 + 2l_1\left\|\Gamma_\phi\right\|\|e\| + 2l_2\|\Gamma_\Omega\|\|e\| \\ - K_{isd}\varepsilon_{isd}^2 - K_{isq}\varepsilon_{isq}^2 \quad (56)$$

where $\left\|\chi_\phi(\varepsilon_\phi)\right\| \leq l_1\|e\|$ and $\|\chi_\Omega(\varepsilon_\Omega)\| \leq l_2\|e\|$; $l_1 > 0, l_2 > 0$.

Considering the following inequalities:

$$\|e\|\left\|\Gamma_\phi\right\| \leq \frac{\zeta_1}{2}\left\|\Gamma_\phi\right\|^2 + \frac{1}{2\zeta_1}\|e\|^2;\ \|e\|\|\Gamma_\Omega\| \leq \frac{\zeta_2}{2}\|\Gamma_\Omega\|^2 + \frac{1}{2\zeta_2}\|e\|^2 \forall \zeta_1, \zeta_2 \in]0, 1[\quad (57)$$

We obtains:

$$\dot{V}_{ic} \leq -\delta_0\|e\|^2 - \eta_\phi\left\|\Gamma_\phi\right\|^2 - \eta_\Omega\|\Gamma_\Omega\|^2 + l_1\zeta_1\left\|\Gamma_\phi\right\|^2 + l_2\zeta_2\|\Gamma_\Omega\|^2 \\ - K_{isd}\varepsilon_{isd}^2 - K_{isq}\varepsilon_{isq}^2 + \frac{l_1}{\zeta_1}\|e\|^2 + \frac{l_2}{\zeta_2}\|e\|^2 \quad (58)$$

Grouping the various terms, we obtain:

$$\dot{V}_{ic} \leq -\left(\delta_0 - \frac{l_1}{\zeta_1} - \frac{l_2}{\zeta_2}\right)\|e\|^2 - (\eta_\phi - l_1\zeta_1)\left\|\Gamma_\phi\right\|^2 - (\eta_\Omega - l_2\zeta_2)\|\Gamma_\Omega\|^2 \\ - 2K_{isd}\left(\frac{1}{2}\varepsilon_{isd}^2\right) - 2K_{isq}\left(\frac{1}{2}\varepsilon_{isq}^2\right) \quad (59)$$

We define $\upsilon = \min(\sigma_1, \sigma_2, \sigma_3, 2K_{isq}, 2K_{isd})$ where $\sigma_1 = \delta_0 - \frac{l_1}{\zeta_1} - \frac{l_2}{\zeta_2}$, $\sigma_2 = \eta_\phi - l_1\zeta_1$, $\sigma_3 = \eta_\Omega - l_2\zeta_2$.

The derivative of V_{ic} becomes:

$$\dot{V}_{ic} \leq -\upsilon V_{oc} \quad (60)$$

By choosing η_ϕ, η_Ω and δ_0 such that σ_1, σ_2 and σ_3 are greater than zero. Then, flux and speed errors converge asymptotically to zero.

5 Simulations Results

These strategies are tested to an induction machine with (see the parameter of the induction machine in Annex) a profile of 7 s.

5.1 Simulations Results Using the Adaptive Observer

Fig. 2 gives the simulation results of the sensorless control for induction machine with nominal parameters: the observer and the controller are designed by using the same parameters that the IM model parameters. Figure 2a shows the estimated speed and the measured one. The speed error is displayed in Fig. 2b. It is to be noted that the estimated speed tracks the actual speed very well.

Now, to illustrate the robustness of the sensorless control scheme, the influence of parameter deviations is investigated. Parameter deviations are intentionally introduced in the observer–controller scheme. First, Fig. 3 is shown the responses for a 50% increase of the stator and rotor resistance. Secondly, Fig. 4 has presented the robustness to the inductances variations.

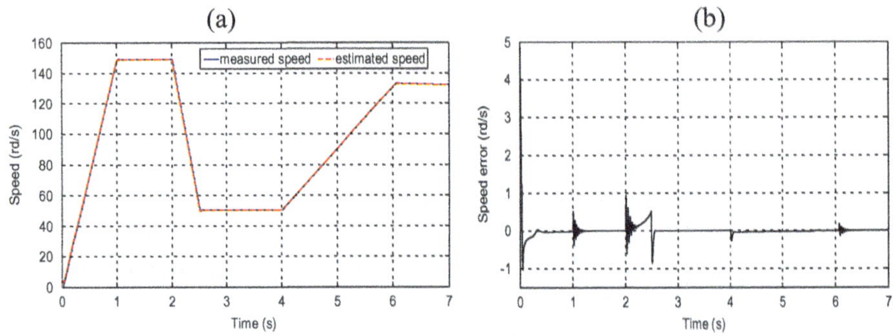

Fig. 2 Speed tracking nominal case: **a**: real and estimated speed, **b** speed error

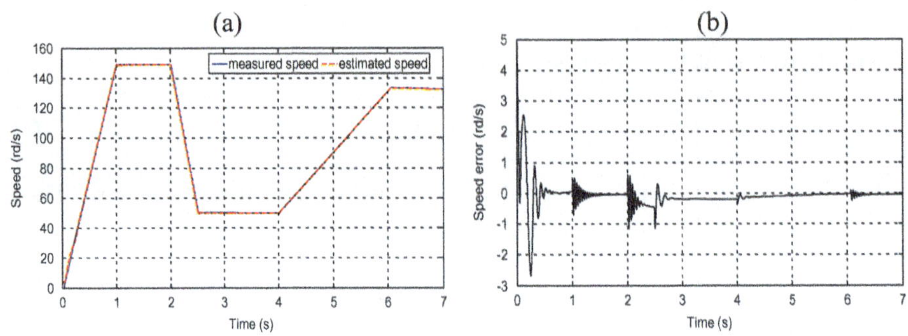

Fig. 3 Speed tracking robustness with respect to + 50% on R_s and R_r : **a**: real and estimated speed, **b** speed error

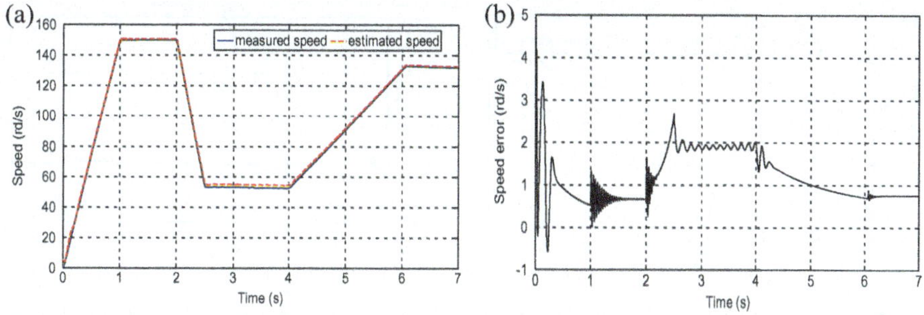

Fig. 4 Speed tracking robustness with respect to + 20% on L_m, L_r and L_s : **a**: real and estimated speed, **b** speed error

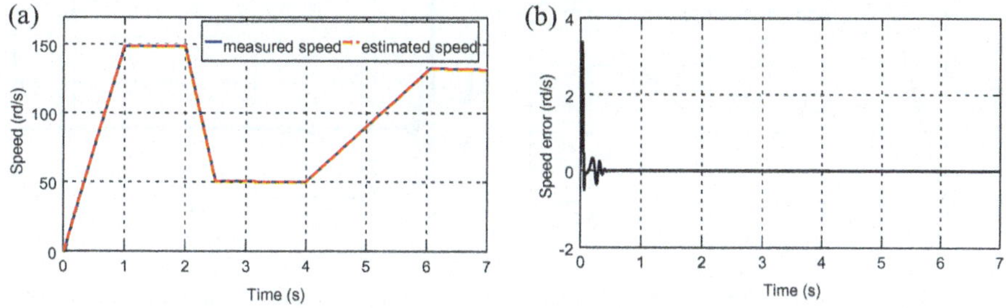

Fig. 5 Speed tracking nominal case: **a**: real and estimated speed, **b** speed error

5.2 Simulations Results Using the Interconnected Observer

The simulation results of the sensorless control for induction machine with nominal parameters are given in Fig. 5. In this case, both observer and controller are tested by using the nominal parameters. Figure 5a, show the estimated magnitudes and the measured ones. The speed error is shown in Fig. 5b, It is remarkable that the estimated speed tracks its actual values very well.

To show the robustness of the suggested sensorless control scheme using the interconnected observer, the impact of parameter deviations is investigated. Parameters variations are included in the observer–controller scheme. First, the response for a 50% increase of the stator and rotor resistance is shown in Fig. 6. Secondly, the robustness to the inductances variations is presented in Fig. 7. According to the error Figs. 6b, 7b, the observer gives desirable results and thus it performs well. The robustness of the suggested sensorless control was so validated.

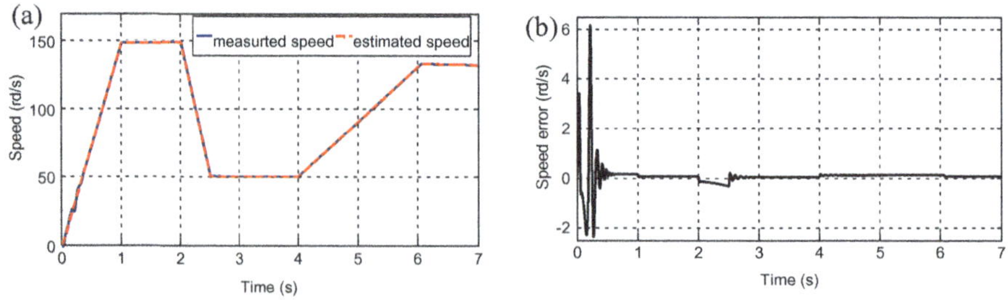

Fig. 6 Speed tracking robustness with respect to $+\,50\%$ on R_s and R_r : **a**: real and estimated speed, **b** speed error

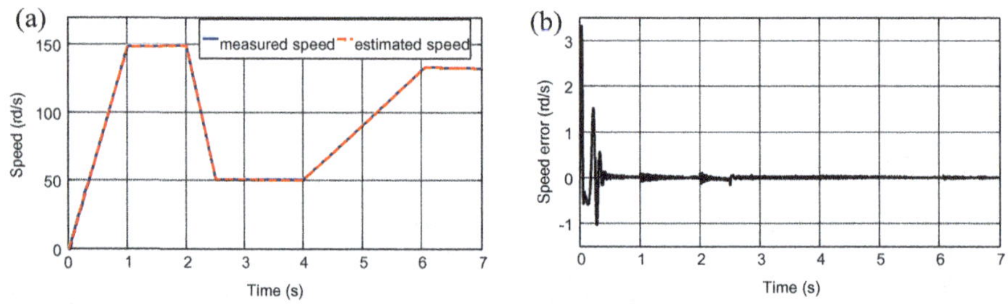

Fig. 7 Speed tracking robustness with respect to -20% on L_m, L_r and L_s : **a**: real and estimated speed, **b** speed error

6 Conclusion

This chapter is concentrated on sensorless control schema for IM combining a robust control with an observer and an interconnected observer. In fact, the proposed control was validated under parametric uncertainties and it performs well. Finally, the global stability of both controller and observer is guaranteed by Lyapunov stability analysis. Simulation results confirm the effectiveness of the proposed methodologies.

References

1. Marino, R., Tomei, P., Verelli, C.M.: A global tracking control for speed sensorless induction motors. Automatica **40**(6), 1071–1077 (2004)
2. Feemster, M., Aquino, P., Dawson, D.M., Behal, A.: Sensorless Rotor velocity tracking control for induction motors. IEEE Trans. Control Syst. Technol. **9**(4), 645–653 (2001)
3. Ghanes, M.: Observation et Commande de la Machine Asynchrone sans Capteur Mécanique, thèse de doctorat, IRCCyN (Ecole Centrale de Nantes) (2005)
4. Traoré, D., de Leon, J., Glumineau, A.: Adaptive interconnected observer-based backstepping control design for sensorless induction motor. Automatica **48**, 682–687 (2012)
5. Chiasson, J.: Non linear controllers For induction motors. In: IFAC Conference System Structure and Control. Nantes, France (1995)

6. Boukettaya, G., et al.: A comparative study of three different sensorless vector control strategies for a flywheel energy storage system. Energy **35**, 132–139 (2010)
7. Naifar, O., Boukettaya, G., Oualha, A., Ouali, A.: A comparative study between a high-gain interconnected observer and an adaptive observer applied to IM-based WECS. Euro. Phys. J. Plus **130**(5), 1–13

The manufacturer's authorised representative in the EU is Springer Nature Customer Service Centre GmbH, Europaplatz 3, 69115 Heidelberg, Germany. If you have any concerns regarding our products, please contact ProductSafety@springernature.com

Printed and bound by CPI Group (UK) Ltd, Croydon, CR0 4YY
30/09/2026
02225332-0001